TISSUE PRINTING

TISSUE PRINTING

TOOLS FOR THE STUDY OF ANATOMY, HISTOCHEMISTRY, AND GENE EXPRESSION

Edited by

Philip D. Reid
Department of Biological Sciences
Smith College
Northampton, Massachusetts

Rafael F. Pont-Lezica
Centre de Physiologie Végétale
Université Paul Sabatier
Toulouse, France

Associate Editors

Elena del Campillo
Department of Plant Biology
University of California at Berkeley
Berkeley, California

Rosannah Taylor
Weed Science Laboratory
Beltsville Agricultural Research Center
Agricultural Research Service
United States Department of Agriculture
Beltsville, Maryland

ACADEMIC PRESS, INC.
Harcourt Brace Jovanovich, Publishers
San Diego   New York   Boston   London
Sydney   Tokyo   Toronto

QH
613
.T57
1992

Academic Press, Inc.
1250 Sixth Avenue, San Diego, California 92101-4311

United Kingdom Edition published by
Academic Press Limited
24–28 Oval Road, London NW1 7DX

Library of Congress Cataloging-in-Publication Data

Tissue printing: tools for the study of anatomy, histochemistry, and
    gene expression / Philip D. Reid . . . [et al.].
        p.   cm.
    Includes bibliographical references and index.
    ISBN 0-12-585970-8
    1. Histochemistry—Methodology.   I. Reid, Philip Dean, date
QH613.T57   1992
574.8'212'028--dc20

92-7412
CIP

PRINTED IN THE UNITED STATES OF AMERICA
92 93 94 95 96 97   BC   9 8 7 6 5 4 3 2 1

This book is dedicated to Professor
Joseph E. Varner, Charles Rebstock Professor
of Biology, Washington University, St. Louis,
Missouri. Teacher, colleague, friend, and
inspiration to us all.

P. D. R., R. P. -L., E. d. C., & R. T.

Contents

CHAPTER 9

Tissue Prints of Animal Tissues
Philip D. Reid

CHAPTER 10

Miscellaneous Applications of Tissue Printing
Rafael F. Pont-Lezica

Contributors

Numbers in parentheses indicate the pages on which the authors' contributions begin.

James D. Anderson, (54), Plant Hormone Laboratory, Beltsville Agricultural Research Center (West), Agricultural Research Service, United States Department of Agriculture, Beltsville, Maryland 20705

Brian A. Bailey, (54), Plant Hormone Laboratory, Beltsville Agricultural Research Center (West), Agricultural Research Service, United States Department of Agriculture, Beltsville, Maryland 20705

Roger N. Beachy, (127), Department of Biology, Washington University, St. Louis, Missouri 63130

Fraulein Cabanag, (89), Department of Chemistry, Silliman University, Dumaguete City 6200, The Philippines

Hilconida Calumpong, (89), Department of Biology, Silliman University, Dumaguete City 6200, The Philippines

Gladys I. Cassab, (23, 63), Instituto de Biotecnología, Universidad Autónoma de Mexico, Cuernavaca, Morelos, 62271 Mexico[1]

John Castelloe, (19), Department of Biology, Washington University, St. Louis, Missouri 63130

Jeffrey F. D. Dean, (54), Plant Hormone Laboratory, Beltsville Agricultural Research Center (West), Agricultural Research Service, United States Department of Agriculture, Beltsville, Maryland 20705

Elena del Campillo, (41), Department of Plant Biology, University of California at Berkeley, Berkeley, California 94720

[1]*Current affiliation*: Department of Plant Biology, University of California at Berkeley, Berkeley, California 94720.

Erwinia S. Duran, (89), Department of Biology, Silliman University, Dumaguete City 6200, The Philippines

William H. Flurkey, (60), Department of Chemistry, Indiana State University, Terre Haute, Indiana 47809

Sue E. Fritz, (32), Department of Biology, Utah State University, Logan Utah 84322

Melissa A. Gee, (100), Department of Biology, University of Missouri, Columbia, Missouri 65211

Tom J. Guilfoyle, (100), Department of Biochemistry, University of Missouri, Columbia, Missouri 65211

Gretchen Hagen, (100), Department of Biochemistry, University of Missouri, Columbia, Missouri 65211

Avtar K. Handa, (47, 121), Department of Horticulture, Purdue University, West Lafayette, Indiana 47907

R. W. F. Harriman, (121), Department of Horticulture, Purdue University, West Lafayette, Indiana 47907

Curtis A. Holt, (125), Department of Biology, Washington University, St. Louis, Missouri 63130[2]

Elizabeth E. Hood, (32), Department of Biology, Utah State University, Logan, Utah 84322

Kendall R. Hood, (32), Department of Biology, Utah State University, Logan, Utah 84322

H. T. Hsu, (131), Florist and Nursery Crops Laboratory, Beltsville Agriculture Research Center, United States Department of Agriculture, Beltsville, Maryland 20705

Y. H. Hsu, (131), Florist and Nursery Crops Laboratory, Beltsville Agriculture Research Center, United States Department of Agriculture, Beltsville, Maryland 20705

Beat Keller, (35), Swiss Federal Research Station for Agronomy Zurich, Postfach 8046, Zurich, Switzerland

J. Paul Knox, (35), Department of Cell Biology, John Innes Institute, Norwich NR4 7UH, United Kingdom

R. Bruce Knox, (155), Department of Cell Biology, John Innes Institute, Norwich NR4 7UH, United Kingdom

J.-J. Lin, (63), Department of Biology, Washington University, St. Louis, Missouri 63130

L.-S. Lin, (63), Department of Biology, Washington University, St. Louis, Missouri 63130

N. S. Lin, (131), Florist and Nursery Crops Laboratory, Beltsville Agriculture Research Center, United States Department of Agriculture, Beltsville, Maryland 20705

[2]*Current affiliation:* Division of Plant Biology, The Scripps Research Institute, La Jolla, California 92037.

Mirasol Magbanua, (89), Department of Biology, Silliman University, Dumaguete City 6200, The Philippines

Bruce A. McClure, (100), Department of Biochemistry, University of Missouri, Columbia, Missouri 65211

Daphne J. Osborne, (65), Department of Plant Sciences, University of Oxford, Oxford OX2 3RA, United Kingdom[3]

Rafael F. Pont-Lezica, (71, 143, 147, 153), Centre de Physiologie Végétale, Université Paul Sabatier, 31062 Toulouse, France

Philip D. Reid, (9, 139), Department of Biological Sciences, Smith College, Northampton, Massachusetts 01630

Keith Roberts, (35), Department of Cell Biology, John-Innes Institute, Norwich NR4 7UH, United Kingdom

Mohan B. Singh, (155), School of Botany, University of Melbourne, Parkville, Victoria 3052, Australia

Yan-Ru Song, (51, 95), Institute of Botany, Academia Sinica, Beijing 100044, China

Nicola Stacey, (35), Department of Cell Biology, John Innes Institute, Norwich NR4 7UH, United Kingdom

Cenk Suphioglu, (155), School of Botany, University of Melbourne, Parkville, Victoria 3052, Australia

Rosannah Taylor, (5, 15, 54, 163, 165), Weed Science Laboratory, Beltsville Agricultural Research Center, Agricultural Research Service, United States Department of Agriculture, Beltsville, Maryland 20705

Denise M. Tieman, (47, 121), Department of Horticulture, Purdue University, West Lafayette, Indiana 47907

Mark L. Tucker, (118), Plant Molecular Biology, United States Department of Agriculture, Beltsville, Maryland 20705

Joseph E. Varner, (1, 15, 19, 51, 59, 63, 95, 163, 165), Department of Biology, Washington University, St. Louis, Missouri 63130

Valerie Vreeland, (89), Department of Plant Biology, University of California at Berkeley, Berkeley, California 94720

Zheng-Hua Ye, (51, 95), Department of Biology, Washington University, St. Louis, Missouri 63130

[3]*Current affiliation:* Oxford Research Unit, Open University, Foxcombe Hall, Boars Hill, Oxford OX1 5HR, United Kingdom.

Preface

The elegant simplicity of the technique of making tissue prints on appropriate substrate materials has resulted in many new applications for both teaching and research, especially following the recent discovery of nitrocellulose as a printing medium. The binding properties of appropriately treated nitrocellulose membranes have made possible the localization of proteins, nucleic acids, and certain carbohydrate moieties in a tissue specific, or in some cases, developmentally specific mode. Some tissues, especially plant tissue that has undergone some lignification, can be used to produce tissue prints that reveal a remarkable amount of anatomical detail without staining. These have been used to permanently record developmental changes over time.

This volume is a collection of protocols detailing uses of tissue prints. Some of these have yet to be fully exploited. We believe that tissue printing will be used increasingly as a technique to study a wide variety of biological problems. The protocols presented here can be easily modified by research biologists or teachers as quick and powerful tools to approach as yet undefined problems. We also believe that tissue printing will gain wide use in studying animal systems, although printing soft tissue presents particular problems.

Whenever possible we have included protocols written by authors of research articles in which tissue printing has been used. It is our belief that having information presented by those who are actively using the method adds both a sense of excitement and a clarity to the protocols as presented here. In some instances it was necessary for the editors to extract protocols from the literature. We have indicated such adaptations in the credit lines for each protocol. We refer the reader to the paper cited for any details that may have resulted from our own omission.

Philip D. Reid
Rafael F. Pont-Lezica

CHAPTER

Introduction

Joseph E. Varner
Department of Biology
Washington University
St. Louis, Missouri

I. Historical Aspects

In 1957 R. Daoust published the first in a series of papers on substrate film printing. This paper and succeeding ones described the localization of protease, amylase, RNase, and DNase by the simple expedient of placing cryostat sections of various organs (liver, kidney, pancreas, and intestine) on substrate films of gelatin, starch, or gelatin–nucleic acid; when the films were incubated for a few minutes and then stained for the substrate, negative images were obtained (Daoust, 1965). The surprisingly sharp images were possible because both the substrates and the enzymes were macromolecules with slow diffusion rates. Examples of the resolutions possible with Daoust's procedures are shown in Figs. 1.1 and 1.2. The subcellular resolution of the α-amylase distribution in barley aleurone layer cells in Fig. 1.3 is possible because

Additional contributions to this chapter have been made by Rosannah Taylor.

1

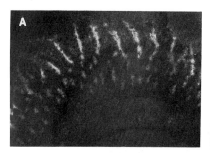
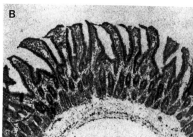

Figure 1.1 (A) A negative image in a gelatin–DNA film exposed to sections of rat ileum for 10 min and stained with toluidine blue (40× magnification). (B) Corresponding tissue section (× 40 magnification) stained with toluidine blue. From "Localization of Deoxyribonuclease in Tissue Sections" by R. Daoust (1957). *Journal of Experimental Cell Research* **12**, 203–211.

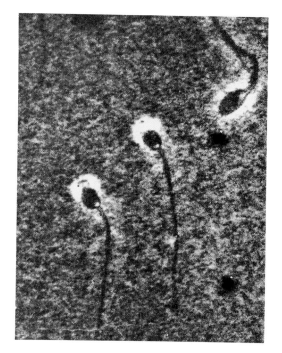

Figure 1.2 Rabbit spermatozoa on gelatin membrane stained with India ink. The light halos result from the proteolytic dispersion of India ink particles (800× magnification). From "Proteolytic Reaction of Mammalian Spermatozoa on Gelatin Membranes" by P. Gaddum and R. J. Blandau (1970). *Science* **170**, 749–751.

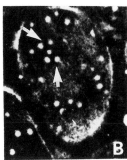

Figure 1.3 α-Amylase in barley aleurone cells is localized by the starch-substrate film method. The light areas, unstained by the starch I₂ complex, show the amylase activity, which is present around the periphery of the cells and around the protein bodies (A: 450× magnification; B: 1230× magnification). From "Cytochemical Localization and Antigenicity of α-Amylase in Barley Aleurone Tissue" by J. V. Jacobsen and R. B. Knox (1973). *Planta* **112**, 213–224.

cells at the developmental stage shown have no large vacuoles (Jacobsen & Knox, 1973).

Several membranes (e.g., nitrocellulose, Nytran, Genescreen, and Immobilon) have been designed to bind proteins and nucleic acids. When the cut surface of a tissue is placed on one of these membranes, the contents of the cut cells transfer to the membrane with little or no lateral movement. The membrane can then be probed for proteins by immunochemistry, for nucleic acids by standard hybridization techniques, and for enzymes by reactions that generate insoluble products.

When a section of plant tissue is pressed onto a nitrocellulose membrane, a physical impression is frequently produced (Fig. 1.4). The impression is a consequence of the relative firmness or hardness of the cell walls. For example, lignified walls make pronounced impressions, and silicified walls actually cut deeply into the nitrocellulose. These physical impressions reveal an astonishing amount of anatomical information when they are viewed with side lighting.

The art and science of substrate film printing, now called *tissue printing*, are still in their early stages; those who set out to make tissue prints will quickly find ways to enhance both their beauty and usefulness.

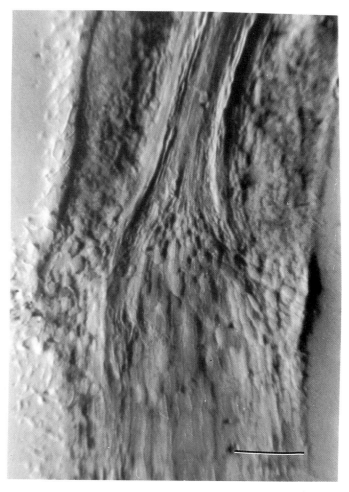

Figure 1.4 Tissue print on nitrocellulose of a longitudinal section of the distal abscission zone of *Phaseolus vulgaris* shows the pulvinus with a central core of vascular tissue (top) and the petiole (bottom). The bar at the bottom of the photograph represents 1 mm. From "Anatomical Changes and Immunolocalization of Cellulase during Abscission as Observed on Nitrocellulose Tissue Prints" by P. D. Reid, E. del Campillo, and L. N. Lewis (1990). *Plant Physiology* **93**, 160–165.

II. Tissue Printing Demonstration[1]

Actually doing tissue printing is the best way to understand its simplicity (Fig. 1.5). Wide variations in the technique are possible and can be developed easily and quickly according to need.

A. Materials

1. Whatman No. 1 filter paper
2. Blotting membrane (nitrocellulose, nylon, PVDF, etc.)
3. Double-edged razor blades
4. Forceps
5. Biological material
6. Rubber gloves
7. Paper to protect the membrane
8. Acrylic sheet
9. Marking pen
10. Hand lens or microscope for viewing the specimen

These materials are shown in Fig. 1.5A.

B. Procedure

1. Place several layers of filter paper on a smooth, hard surface, and place a blotting membrane on top. Use a double-edged razor blade to cut a tissue section 0.2–2 mm thick, depending on the particular tissue sample. It may be necessary to gently preblot the section on a separate piece of membrane or filter paper before printing to remove excess tissue exudate from cut cells and to ensure a faithful print (Fig. 1.5B).
2. Using forceps, transfer the tissue section to the membrane, being careful not to move the section after it is in contact so as to avoid smearing. Several successive sections can be printed on the same piece of membrane (Fig. 1.5C).

[1]This section was contributed by Rosannah Taylor.

Figure 1.5 Detailed descriptions of the steps involved in tissue printing and the materials required are given in text.

3. Place a small piece of nonabsorbent paper over the section to protect the membrane from fingerprints. In some instances rubber gloves may be required. When printing a thin section (200–300 μm), place a piece of membrane on top of the section to prevent the nonabsorbent paper from marking the membrane under the section (Fig. 1.5D).
4. Apply the appropriate amount of pressure to the section for the type of print desired. A chemical print requires only light pressure, but a physical print requires several times as much. The proper pressure also varies with the tissue used (Fig. 1.5E).
5. Gently remove the protective paper and the section with forceps, air dry the print with warm air, and observe (Fig. 1.5F).

Prints may be illuminated from the top or from one side by white or ultraviolet light and may also be viewed with transmitted light.

CHAPTER 2

Physical Tissue Prints

Philip D. Reid

Department of Biological Sciences
Smith College
Northampton, Massachusetts

I. Overview
II. Stem Segments and Abscission Zones of Cotton
III. Physical Printing to Study the Role of Calcium in
   Maintaining Cell Wall Firmness
IV. Visualization of Epidermal Glandular Structures in Plants
V. High-Resolution Tissue Prints on Agarose

I. Overview

Segments of freshly sectioned tissue pressed into nitrocellulose membranes leave imprints that reveal anatomical detail that is otherwise difficult to see without fixing, embedding, and then sectioning the

Additional contributions to this chapter have been made by Rosannah Taylor, Joseph E. Varner, and John Castelloe.

9

segments. Physical tissue prints are easily made by this technique and are useful for studying tissue organization and developmental changes that take place over time. Although the technique works for nearly any plant segment, tissues that are very soft, such as those of *Impatiens*, do not print as well as tissues that contain at least some lignified cells. Segments of bean, cotton and many other plants make excellent tissue prints for teaching and research applications.

Plant tissue imprints on nitrocellulose membranes were first described in 1987 by Cassab and Varner and by Spruce, Mayer, and Osborne. These researchers looked for specific proteins by developing the tissue prints with specific histochemical reagents. The initial results were extended to demonstrate the use of physical prints for studying abscission in bean petioles and to demonstrate that abscission-specific enzymes could be localized on the physical tissue prints (Reid & del Campillo, 1989).

Physical tissue prints can be observed with a low-power lens by transmitted light. When viewed this way, lignified cells appear bright and show a remarkable amount of anatomical detail. Most cells in a typical stem cross section can be identified by this method, although the conducting cells in the phloem have not been clearly observed in any section, probably because they are too soft to leave a distinct imprint on the membrane. If the light source is changed so that it is oblique to the microscope stage (e.g., a Nicholas illuminator placed at the side of the microscope), the print appears as a three-dimensional image and additional anatomical detail is revealed. Unlabeled prints have shown that tissue integrity changes during abscission in bean petioles (Reid, del Campillo, & Lewis, 1990).

The resolution of anatomical prints is inferior to that of fixed and stained sections, but it is adequate to resolve annular thickenings on primary xylem cell walls. Unfortunately, when tissue prints are developed by exposure to histochemical labeling treatments, their resolution becomes poorer. On the other hand, the high-resolution prints on agarose described in Section V demonstrate that refining the process for tissue printing leads to improvement in the resolution of labeled prints.

II. Stem Segments and Abscission Zones of Cotton

Four-week-old seedlings of *Gossypium hirsutum* (cotton) provide excellent material for physical tissue printing. The stems and petioles of most cultivars contain glands that include various phenolic substances, such as gossypol, that leave a distinct red stain on the print. A ring of vascular tissue clearly shows the lignified xylem cells and phloem fibers in Fig. 2.1A, a photograph taken with light transmitted through the nitrocellulose tissue print. Figure 2.1B shows a longitudinal view of the same stem segment. Figure 2.2 demonstrates that cellular detail, such as annular thickenings on primary xylem elements, can be observed on the tissue prints. Figure 2.3, which shows a side-illuminated cotton node that has begun to abscise, illustrates the use of tissue printing to study a developmental process and reveals the three-dimensional aspect that results from shadows cast by projections on the tissue print.

A. Materials

1. Nitrocellulose membrane, 0.45-μm pore size, cut to fit a microscope slide
2. Protective paper: glassine paper or paper that manufacturers place between nitrocellulose sheets
3. Paper towels
4. Cotton stem segment (see Fig 2.3): Grow *Gossypium hirsutum* seeds for 4 wk in plant growth chamber; take stem segment

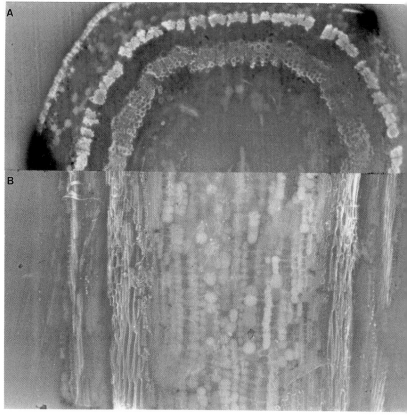

Figure 2.1 (A) Cross section and (B) longitudinal section of a stem at the first internode of a 4-wk-old cotton plant are printed onto nitrocellulose and observed with transmitted light (2.5× magnification).

from oldest internode, between cotyledonary node and that of first true leaf; induce abscission by treating debladed cotton explant with 25 µl/liter ethylene gas for 72 hr

B. Procedure

1. Place dry nitrocellulose, smooth side up, on two layers of paper towel lying on a glass surface or smooth laboratory bench.

Figure 2.2 Longitudinal section of cotton stem vascular tissue shows spiral thickenings on the walls of primary xylem cells (25× magnification).

2. Section with a sharp razor blade.
3. Gently place the cut surface of the tissue to be printed on the nitrocellulose membrane.
4. Cover the section with protective paper.
5. Press the section onto the membrane firmly with your finger. The proper amount of pressure varies, but trial and error will soon indicate a pressure that yields good results.
6. Remove the protective paper.
7. Using surgical forceps, carefully remove the tissue section from the nitrocellulose membrane.
8. Attach the membrane to a microscope slide, and observe with either transmitted light or light from the side of the microscope stage.

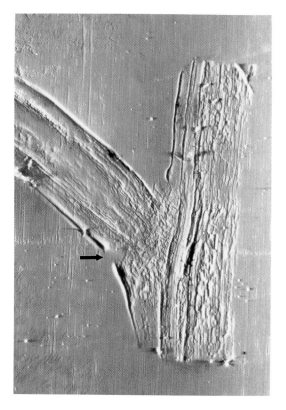

Figure 2.3 Side illumination reveals three-dimensional aspect of a longitudinal section through a cotton node printed onto nitrocellulose after abscission has been induced by ethylene (2.5× magnification).

Notes

1. Thin sections work best with soft tissues, such as bean (*Phaseolus*) tissues. Harder tissues, such as those of cotton (*Gossypium*), are resistant to deformation by pressure, so tissue thickness is less critical. Sections of soft tissue can be easily prepared by holding the segment to be cut between microscope slides (taped together on either side of the segment). With practice, sections about 200 µm thick can be routinely prepared this way.

2. The light for transmission through the nitrocellulose membrane needs to be reasonably bright to illuminate the lignified cells of the tissue segment. Very bright light is required for the best resolution.

III. Physical Printing to Study the Role of Calcium in Maintaining Cell Wall Firmness[1]

A variety of techniques have been used to demonstrate the function of calcium in plant cells. However, tissue printing may also be used to show how calcium affects cell wall firmness. The use of tissue printing to indicate physical anatomy on nitrocellulose membranes was first described by Cassab and Varner in 1987, and previously we reported on the relationship of Ca^{2+} to cell wall firmness by using tissue printing on nitrocellulose (Varner and Taylor 1989).

This exercise uses tissue printing to demonstrate the role of calcium in maintaining cell wall firmness in soybean petioles; treatment of sections with EGTA softens the walls of the epidermis, phloem, and primary and secondary xylem so that they do not make a physical print on nitrocellulose membranes (Fig. 2.4).

A. Materials

1. Soybean seedlings cultivated in the greenhouse with natural-day-length illumination
2. Nitrocellulose membrane, 0.45-μm pore size (Schleicher & Schuell type BA-85)
3. 10 mM $CaCl_2$ in 100 mM sodium acetate (pH 5.6)
4. 100 mM EGTA in 100 mM sodium acetate (pH 5.6)
5. Staining dishes

[1]This section was contributed by Rosannah Taylor and Joseph E. Varner.

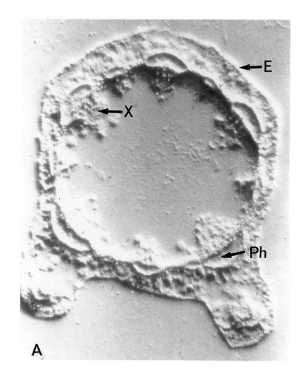

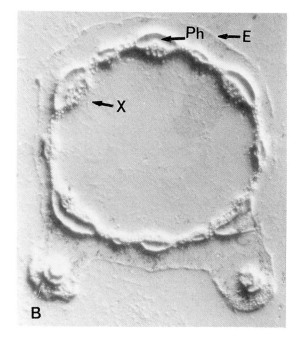

B. Procedure

1. Make freehand sections (1 or 2 mm thick) of soybean petiole, and place half of the sections into 10 mM CaCl$_2$ in 100 mM sodium acetate (pH 5.6) and the other half into 100 mM EGTA in 100 mM sodium acetate (pH 5.6). Treat for 3 hr.
2. After 3 hr, select sections from each of the treatments for physical printing.
3. Rinse the sections briefly in distilled water.
4. Preblot each section on a separate piece of membrane or filter paper before printing.
5. Make physical prints by placing each section on an untreated nitrocellulose membrane. Place a piece of smooth paper over the section to protect the membrane from fingerprints, and then apply firm pressure on the tissue for 10–30 sec without crushing.
6. Dry and observe.
7. Photograph the prints with unilateral side lighting. Be sure the angle of the light is the same for all of the sections.

IV. Visualization of Epidermal Glandular Structures in Plants[2]

The distribution of microscopic glandular structures (hairs, capsules, and nodules) on a leaf or other plant part can be visually recorded by

[2]This section was adapted from "Techniques for Visualization of Epidermal Glandular Structures in Plants" T. Eisner, M. Eisner, and J. Meinwald (1987). *Journal of Chemical Ecology* **13**, 943–946.

Figure 2.4 (A) Freehand sections of soybean petioles are treated with 10 mM CaCl$_2$ in 100 mM sodium acetate (pH 5.6) for 3 hr, and physical impressions of the epidermis (E), phloem (Ph), and primary and secondary xylem (x) are obtained. (B) Treatment of sections with 100 mM EGTA in 100 mM sodium acetate (pH 5.6) for 3 hr softens the walls of the epidermis, phloem, and primary and secondary xylem so they do not make a physical print on nitrocellulose membranes. These data show that Ca^{2+} is required for maintaining firmness of some cell walls.

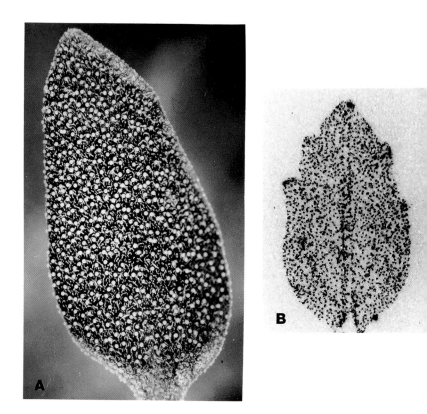

Figure 2.5 (A) Side-illuminated leaf of *Teucrium marum* (cat thyme) shows the distribution of stalkless glandular capsules. (B) This distribution can be visually recorded with a tissue print of the leaf on an indicator surface soaked in Tollens reagent. From "Technique for Visualization of Epidermal Glandular Structures in Plants" by T. Eisner, M. Eisner, and J. Meinwald (1987). *Journal of Chemical Ecology* **13**, 943–946.

pressing the plant part against an indicator surface (filter paper or thin-layer chromatography plate) soaked in Tollens reagent. The glandular sites are revealed instantly as dark spots on the test surface, as shown in Fig. 2.5 (Eisner, Eisner, & Meinwald, 1987).

A. Materials

1. Plant leaves or stems containing microscopic glandular structures, such as secretory hairs, capsules, or nodules

2. Indicator surfaces, such as filter paper or thin-layer silica-gel chromatography plates
3. Tollens reagent (Tollens, 1882a, 1882b; Feigl, 1966) made by adding droplets of aqueous potassium hydroxide (10%) to aqueous $AgNO^3$ (10%) and dissolving the resulting silver oxide precipitate by adding aqueous ammonia (10%)

B. Procedure

1. Soak filter paper or silica-gel plates in freshly prepared Tollens reagent just before using.
2. Blot to remove excess reagent.
3. Press the plant tissue tightly against the indicator surface to rupture the glandular capsules and nodules. Glandular sites are immediately revealed as a pattern of dark spots on the indicator surface.
4. Observe the prints under low-power magnification.

Note

Tollens reagent deteriorates over time and is potentially explosive, but a fresh preparation from the stable precursors poses no problems.

V. High-Resolution Tissue Prints on Agarose[3]

When a section of plant tissue is pressed onto any plastic surface, a more or less faithful impression is obtained. The hardest cell walls, that is, lignified, silicified, and cutinized walls, leave impressions on nitrocellulose membrane, on the dry emulsion of 35-mm black-and-white photographic film, or even on the cellulose acetate side of 35-mm film. However, a much softer printing medium is required for obtaining impressions of large cells with unhardened walls. We have found that 6% agarose takes high-resolution impressions (Fig. 2.6). To keep the agarose film receptive for days to weeks, we add sorbitol and glycerol

---

[3]This section was contributed by John Castelloe and Joseph E. Varner.

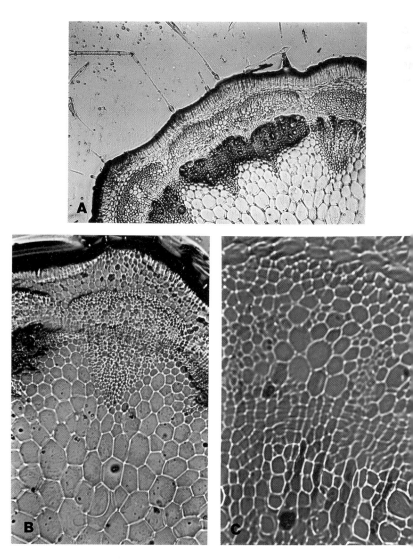

Figure 2.6 Agarose (6% agarose, 15% sorbitol) provided this high-resolution impression of a freehand cross section of a soybean stem, shown at (A) 65× magnification, (B) 115×, and (C) 455×. The impression of the cambial cells is blurred because the walls of these cells are less firm and the impression lies on a plane different from that of the impressions of the xylem and phloem cells.

to final concentrations of 10% and 5%, respectively. This procedure was inspired by and derived from a research paper by Howlett, Knox, and Heslop-Harrison (1973) on the use of agarose films to study the release of pollen antigens.

A. Materials

1. Agarose (gelling temperature ≈36°C) (Sigma)
2. Sorbitol (Sigma)
3. Glycerol (Sigma)
4. Microscope slides (frosted), 3 in. × 1 in. × 1 mm (Fisher)

B. Preparation of Agarose-Coated Slides

1. Place 3 g of agarose, 2.5 g of glycerol, and 5 g of sorbitol in a 125-ml Erlenmeyer flask; add enough distilled water to bring the final volume to 50 ml.
2. Warm and stir until all components are dissolved; continue heating at just under the boiling temperature until most of the air bubbles are gone.
3. Pour a few drops on a microscope slide; tilt the slide to let the solution cover about two-thirds of the upper surface, and then hold the slide in a vertical position to let the excess solution run off.
4. The thickness of the coating on the slide may be varied by (a) letting the solution stand at room temperature for several seconds to several minutes before pouring; (b) keeping the slide horizontal for different amounts of time before letting the solution run off; (c) preheating the slide to various temperatures; or (d) varying the concentrations of agarose, sorbitol, and glycerol.
5. Film coatings made by this formula take the best impressions if they are allowed to stand for a few days before use.

C. Procedure

1. Make a ≈0.5 mm-thick section (either cross or longitudinal) with a sharp razor blade.
2. Place the section on the coated slide.
3. Place an appropriately cut piece of Scotch tape over the section. Press the tape with a fingertip, using a slight back-and-forth and side-to-side motion to distribute pressure evenly on all areas of the section.
4. Remove the section by pulling it away with the Scotch tape. The impression is stable for at least several weeks.

Note

Tissue prints on agarose are most readily seen under a microscope with dark-field illumination (Fig. 2.6).

CHAPTER 3

Localization of Cell Wall Proteins

Gladys I. Cassab[1]

Instituto de Biotecnología
Universidad Autónoma de Mexico
Cuernavaca, Morelos, Mexico

I. Overview
II. Localization of Extensin in Soybean Seeds and Hypocotyls
III. Localization of Extensin in Maize Cell Walls Using
Monoclonal Antibodies
IV. Localization of Glycine-Rich Protein 1.8 in Bean Ovaries
and Hypocotyls
V. Tissue Printing to Screen for Mutants: A Search for Cell Surface
Carbohydrate Mutants in *Arabidopsis*

I. Overview

The presence of walls is one of the outstanding characteristics distin-
guishing the cells of plants from those of animals. The cell wall is complex and has unique characteristics related to the developmental stage

Additional contributions to this chapter have been made by Elizabeth E. Hood,
Kendall R. Hood, and Sue E. Fritz, and Beat Keller, Nicola Stacey, J. Paul Knox, and
Keith Roberts.

[1]*Current affiliation:* Department of Plant Biology, University of California at
Berkeley, Berkeley, California.

23

of a given plant cell type. The cell wall is not, like the mitochondrion, an organelle with relatively constant duties; rather, it is subject to the continuous developmental processes that govern cell size, division, shape, and function (Cassab & Varner, 1988).

Cell walls are composed of cellulose, hemicellulose, pectic compounds, lignin, suberin, proteins, and water. In addition, they may contain structural proteins and enzymes. The best-characterized and perhaps most abundant structural proteins of dicotyledon cell walls are the extensins, members of a class of hydroxyproline-rich glycoproteins (HRGPs) present in a wide variety of plants and algae (Cassab & Varner, 1988; Varner & Lin, 1989). Extensins are basic proteins with a high content of hydroxyproline (36–42%) that typically are glycosylated with one to four arabinosyl residues. They contain repeating amino acid sequences and assume a polyproline-II helical structure. Chen and Varner isolated a complementary DNA (cDNA) clone for carrot extensin (1985a), and they also characterized a carrot extensin genomic clone (1985b). The derived amino acid sequence from the genomic clone contained a putative signal peptide and 25 Ser-(Pro)$_4$ sequences. Two different extensin RNA transcripts were found corresponding to the genomic clone with different 5′ start sites. Both transcripts increased markedly after wounding, which correlates with the extensin accumulation seen in the cell walls of carrot roots after wounding (Chen & Varner, 1985b). It was once thought that extensins were the major protein component of the primary wall of all plant cells and that they had a function in cell wall architecture (Lamport, 1980). However, cellular biology has demonstrated that extensin is most abundantly localized in sclerenchyma tissue cells (Cassab & Varner, 1987). Sclerenchyma cells act as the skeletal elements of the plant body, enabling it to withstand mechanical stresses, such as bending, compression, and tension (Esau, 1965; Haberlandt, 1914). Thus, the presence of extensin in the sclerenchyma cell walls, together with other wall components, may determine the unique characteristics of sclerenchyma cells.

Studying extensin localization in plant cells by using anti-soybean extensin antibodies and tissue printing on nitrocellulose paper (Cassab & Varner, 1987) has helped us to understand the action of extensin in plants. Extensin is a difficult protein to isolate, because a high proportion of it becomes insolubilized in the wall compartment. The mechanism of its insolubilization has not yet been elucidated (Cassab &

Varner, 1988), and the *in muro* interactions of extensin with other cell wall components are not understood. Tissue printing on nitrocellulose paper is a new approach for analyzing the distribution of extensin in different types of plant cells. This technique involves a simple immunolocalization procedure that should be generally useful for studying cell wall proteins.

Little is known about the cell wall proteins of monocotyledons. However, an extensinlike molecule has been characterized in the walls of maize cells grown in suspension culture (Kieliszewski & Lamport, 1987), in the developing maize pericarp (Hood, Shen, & Varner, 1988), and in several embryonic and postembryonic maize tissues (Stiefel *et al.*, 1990; Ruiz-Avila, Ludevid, & Puigdomenench, 1991). Also, a cDNA clone has been isolated from maize coleoptiles (Stiefel *et al.*, 1988). The deduced amino acid sequence from this cDNA corresponds to the hydroxyproline- and threonine-rich glycoprotein extracted from cell walls of suspension culture cells and various seedlings and embryo tissues (Kieliszewski & Lamport, 1987; Kieliszewski, Leykam, & Lamport, 1990) and of maize pericarps (Hood, Shen, & Varner, 1988). Extensins from maize have been immunolocalized in the walls of root tip cells by using specific antibodies against the purified maize protein (Ludevid *et al.*, 1990). The accumulation of mRNA corresponding to this protein has been analyzed by using the cDNA probe. The expression of this gene has been found in tissues with high proportions of dividing cells, such as the root tip of young maize seedlings. The maize mRNA for extensin has been induced by wounding young leaves and coleoptiles (Ludevid *et al.*, 1990). Moreover, the spatial pattern of expression of extensin from maize has been analyzed by *in situ* hybridization. During normal development of roots and leaves, the expression of the gene is transient and high in regions initiating vascular elements and associated sclerenchyma. Its expression is also associated with the differentiation of vascular elements in a variety of other tissues (Stiefel *et al.*, 1990).

Extensin is not the only structural protein of the cell wall. Cell wall proteins unrelated to extensin have been shown to accumulate in response to a variety of developmental and stress signals (Bozart & Boyer, 1987; Franssen, Nap, Gloudemans, Stiekema, & Van Dam, 1987; Tierney, Weichert, & Pluymers, 1988; Datta, Schmidt, & Marcus, 1989; Hong, Nagao, & Key, 1989; Keller & Lamb, 1989). Some plant cells are rich in glycine (Varner & Cassab, 1986). A petunia gene encoding a protein that is 67% glycine has been characterized (Condit & Meagher, 1987). Two similar genes occur in *Phaseolus vulgaris* (Keller, Sauer, & Lamb, 1988); one of the encoded glycine-rich proteins (GRP 1.8) was localized by tissue printing and immunocytochemical studies of the cell walls in vascular tissue of young bean ovaries and hypocotyls (Keller, Templeton, & Lamb, 1989). The cell walls of barley leaves contain abundant polypeptides, called thionins, with antifungal activity (Bohlmann et al., 1988). Thionins have molecular weights of about 5000, are basic, are rich in cysteine (17%), and constitute the major soluble protein of barley leaf cell walls.

The arabinogalactan proteins (AGPs), members of the HRGP class, are widely distributed in plants. They are primarily localized in the extracellular matrix and in gums and exudates (Fincher, Stone, & Clarke, 1983). The AGP associated with cell walls is freely soluble in water and is thought to have a cell-to-cell recognition, rather than structural, function. The styles of flowers frequently contain arabino-galactan, possibly associated with protein (Hoggart & Clarke, 1984), and the medullae of soybean root nodules are rich in AGP (Cassab, 1986). The characterization of monoclonal antibodies to AGPs (Pennel, Knox, Scofield, Selvedran, & Roberts, 1989; Knox, Day, & Robertson, 1989) has been useful for studying the possible function of AGPs in plants. The expression of the AGP epitopes in plant cells shows strict developmental regulation (Pennel & Roberts, 1989; Knox, Day, & Roberts, 1989; Stacey, Roberts, & Knox, 1990). A protocol using tissue printing for screening AGP mutants is provided in Section V. The precise function of AGPs is unknown, but their observed patterns of expression, their location at the plasma membrane, and their known ability to react with Yariv antigen (Fincher, Stone, & Clarke, 1983) may indicate that they function in molecular recognition and cell-to-cell interaction in relation to cell identity or position (Knox, 1990). The localization of peroxidase activity in the cell walls of vascular bundles by tissue printing was reported by Cassab, Lin, Lin, and Varner in 1988.

Several enzymes are associated with cell walls. These include peroxidases, phosphatases, β-1,3-glucanases, β-1,4-glucanases, poly-galacturonase, pectin methylesterases, malate dehydrogenases, β-glucuronidases, β-xylosidases, proteases, and ascorbic acid oxidase (Cassab & Varner, 1988; Varner & Lin, 1989). Tissue printing has been used to localize peroxidase activity in the cell walls of the vascular bundles in pea epicotyls (Cassab, Lin, Lin, & Varner, 1988).

II. Localization of Extensin in Soybean Seeds and Hypocotyls

Nitrocellulose membrane adsorbs relatively large quantities of proteins that are tightly bound (Kuno & Kihara, 1967), whereas it usually does not retain salts, many small molecules, or RNA. Nitrocellulose paper has been successfully used for transferring proteins subjected to polyacrylamide gel electrophoresis in Western blot analysis (Towbin, Staehlin, & Gordon, 1986). The principle of the tissue print Western blot is based on the fact that blotting a tissue section onto nitrocellulose paper leaves a stable faithful image of the cut surface. Because extensins are usually extracted with a solution that has a high salt concentration (Cassab, Nieto-Sotelo, Cooper, Van Holst, & Varner, 1985), I assumed that cell wall proteins would transfer to nitrocellulose paper previously soaked in 0.2 M $CaCl_2$; this transfer does, in fact, occur. The simple procedure (Cassab & Varner, 1987) that follows allows the immunolocalization of cell wall proteins, and the results correlate with those of conventional light microscopy methods (Fig. 3.1).

In soybean seeds, extensin is primarily localized in the seed coat, hilum, and vascular bundles of the cotyledon. In soybean hypocotyls, however, the pattern of extensin distribution varies according to the developmental stage of the tissue and the tissue type (Cassab, Lin, Lin, & Varner, 1988). From the apical hook throughout the elongating region of the hypocotyl, extensin is localized in the cortex and pith parenchyma. When the mature region is printed, extensin is not detected in either the cortex or the pith parenchyma, but it is observed in the vascular bundles. This same pattern occurs in a cross section of the mature root (Fig. 3.2).

As new structural cell wall proteins are discovered, it is likely that preparation of specific antibodies for these proteins will continue to

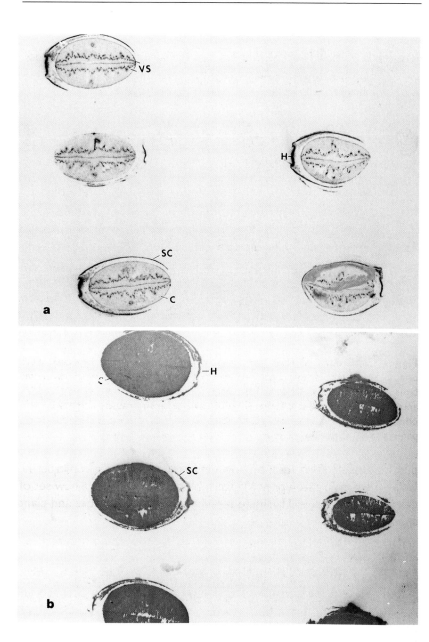

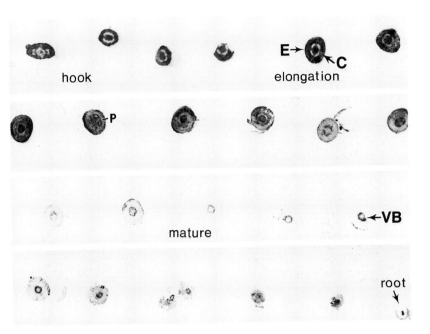

Figure 3.2 Cross sections (≈2 mm thick) of a 3-day-old soybean hypocotyl were printed onto nitrocellulose paper sequentially from the tip of the hypocotyl to the boundary between the shoot and the root. The prints were reacted with specific anti-extensin antibodies diluted 1:15,000, and extensin was detected with AP-conjugated anti-rabbit goat IgG. Abbreviations: C, cortex; E, epidermis; P, parenchyma; VB, vascular bundles.

provide useful information on how different cell walls are constructed. Using the simple tissue print Western blot technique with a new set of cell wall antibodies will facilitate screening many plant tissues and plant species.

Figure 3.1 Soybean seed prints were developed on nitrocellulose paper by using cross sections 3 mm thick. (a) The prints were reacted with specific anti-extensin antibodies diluted 1:15,000, and extensin was detected with AP-conjugated anti-rabbit goat IgG. (b) Prints stained with India ink produce similar results. Abbreviations: C, cotyledon; H, hilum; SC, seed coat; VS, vascular supply of the seed. From "Immunocytolocalization of Extensin in Developing Soybean Seed Coats by Immunogold–Silver Staining and by Tissue Printing on Nitrocellulose Paper" by G. I. Cassab and J. E. Varner (1987). *Journal of Cell Biology* **105**, 2581–2588.

A. Materials

1. Nitrocellulose membrane (Schleicher & Schuell type BA-85)
2. Specific anti-extensin polyclonal antibodies raised from purified soybean seed coat extensin (Cassab & Varner, 1987); anti-extensin antibody diluted 1:15,000
3. Alkaline phosphatase (AP) conjugated anti-rabbit immunoglobulin IgG (Fc)
4. 0.2 M $CaCl_2 \cdot 2H_2O$
5. Tris buffer saline (TBS): 0.9% NaCl in 20 mM Tris-HCl (pH 7.4) plus 0.3% (v/v) Tween-20 and 0.05% NaN[3]
6. Antibody incubation solution: 0.25% (w/v) bovine serum albumin (BSA), 0.25% (w/v) gelatin, and 0.3% (v/v) Tween-20 in TBS
7. AP buffer: 50 mM Tris-HCl (pH 9.8) plus 1 mM $MgCl_2$
8. AP substrate solution: 66 μl nitro blue tetrazolium (NBT) and 33 μl 5-bromo-4-chloro-3-indolyl phosphate (BCIP) (Promega) in 10 ml of AP buffer
9. India ink
10. Kodak Technical Pan 2415 film, ISO 50.

B. Procedure

1. Soak the nitrocellulose membrane in 0.2 M $CaCl_2 \cdot 2H_2O$ for 30 min, and dry on 3-mm Whatman paper.
2. Cut fresh tissue into sections 0.3–3 mm thick with a new razor blade that has been washed in distilled water for 3 sec and dried on Kimwipes. Then, pick up each section with forceps, and carefully place it on the nitrocellulose membrane. Using a gloved fingertip, press the tissue onto the nitrocellulose membrane for 15–30 sec. Finally, carefully remove the tissue section with forceps, and immediately dry the tissue print with warm air.
3. The nitrocellulose membrane is treated as described for protein Western blots by Blake, Johnston, Russell-Jones, and Gotschlich (1981). The first step is to block the nitrocellulose with antibody incubation solution for 1–3 hr at room temperature with constant shaking.

4. Add the primary antibody to the desired dilution in antibody incubation solution, and then incubate the nitrocellulose membrane 1–3 hr at room temperature with constant shaking.
5. After antibody incubation, wash the nitrocellulose membrane three times for 30 min each in TBS with agitation.
6. Soak the nitrocellulose membrane in AP-conjugated anti-IgG diluted 1:20,000 in antibody incubation solution for 1–3 hr with agitation.
7. Wash the nitrocellulose membrane three times for 30 min each in TBS with agitation.
8. Briefly wash the nitrocellulose membrane with AP buffer, and then add the AP substrate solution.
9. Develop the tissue print until a color signal appears, and stop the reaction by washing the nitrocellulose paper quickly in distilled water.
10. Dry the nitrocellulose paper on 3-mm Whatman paper.
11. Photograph the tissue prints with Kodak Technical Pan 2415 film.

For determining the total protein distribution pattern of the tissue section, follow the procedure for tissue printing just described. Then, incubate the nitrocellulose paper 15 min with 1 µl/ml India ink in TBS with constant shaking (Hancock & Tsang, 1983). Afterward, wash the nitrocellulose paper three times in TBS, 5 min each, and air dry on 3-mm Whatman paper.

Notes

1. All instructions for making the tissue print should be strictly followed because double images will be obtained if the tissue section is not carefully blotted and removed from the nitrocellulose filter.
2. Detection of the AP-conjugated second antibody on tissue prints was selected over the peroxidase-conjugated second antibody procedure because the substrates used for detecting the peroxidase, such as O-phenylenediamine and H_2O_2,

are capable of detecting endogenous peroxidase activity in plant tissue sections (Cassab, Lin, Lin, & Varner, 1988). Thus, the endogenous peroxidase activity in the tissue print would mask the immunoblotting reaction.

III. Localization of Extensin in Maize Cell Walls Using Monoclonal Antibodies[2]

The plant cell wall is dynamic and functions in plant structure, cell-to-cell recognition, and resistance to pathogens. It is composed of complex polysaccharides and proteins, with the HRGPs constituting a major portion of the protein component of this matrix. To help determine the role of the HRGPs in the structure of the extracellular matrix, we used tissue prints to assess their distribution in various cell types in maize tissues (Fritz et al., 1991). We found PC-1, the major maize HRGP, in pericarp, vascular bundles, and epidermis tissues that primarily protect and support the plant (Hood et al., 1988). Interestingly, we found the highest amounts of PC-1 in floral silks, tissues that require both strength and flexibility (Hood et al., 1991). Following is a description of the procedure used in these investigations, and the results obtained for developing kernels and for stem and silk are shown in Figs. 3.3 and 3.4, respectively.

A. Materials

1. Nitrocellulose membranes (Biotrace NT, Gelman)
2. Tween Tris buffer saline (TTBS): 150 mM NaCl, 10 mM Tris-HCl (pH 8), and 0.05% Tween-20
3. Primary antibody: anti-PC-1 monoclonal antibody diluted 1:10,000 in TTBS
4. Secondary antibody: AP-conjugated anti-mouse polyvalent

[2]This section was contributed by Elizabeth E. Hood, Kendall R. Hood, and Sue E. Fritz.

| | 0 | 5 | 10 | 20 | 30 | 40 |

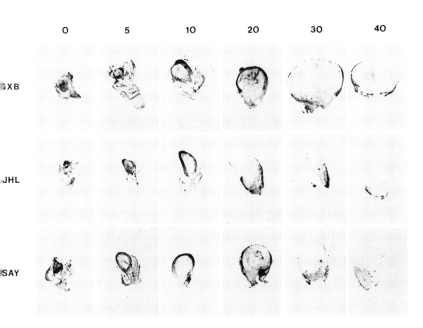

GXB

JHL

SAY

Figure 3.3 Immunostained tissue prints were made with developing kernels from three maize varieties: Golden × Bantam (G × B), a hybrid sweet corn, and Japanese Hulless (JHL) and South American Yellow (SAY), two popcorn varieties. The prints were treated with a 1:10,000 dilution of an anti-PC-1 monoclonal antibody. The secondary antibody was a 1:10,000 dilution of AP-conjugated goat anti-mouse polyvalent. Color was developed with NBT and BCIP. The numbers across the top are days after pollination.

immunoglobulin, NBT (50 mg/ml in 70% methanol), and BCIP (50 mg/ml in 100% dimethylformamide) (Sigma) diluted 1:10,000 in TTBS

5. AP buffer: 100 mM Tris-HCl (pH 9.5), 100 mM NaCl, and 5 mM MgCl^2

6. Antibody incubation solution: 3% BSA in TTBS

7. Color reagent: 66 μl NBT and 33 μl BCIP in 10 ml AP buffer

8. 0.2 M CaCl$_2$

9. Fresh whole maize plants

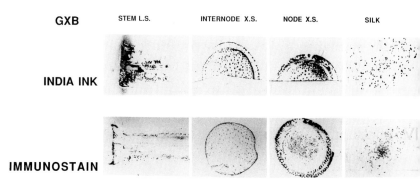

GXB STEM L.S. INTERNODE X.S. NODE X.S. SILK

INDIA INK

IMMUNOSTAIN

Figure 3.4 Tissue prints of longitudinal sections (LS) and cross sections (XS) of stem and silk from *Zea mays* L., Golden × Bantam variety, were either stained for total protein with India ink (Hancock & Tsang, 1983) or reacted with an anti- PC-I monoclonal antibody as described in Fig. 3.3.

B. Procedure

1. Soak the nitrocellulose membranes in 0.2 M CaCl$_2$ for 5 min and dry before printing tissues.
2. Prepare the fresh vegetative tissues for printing by washing with tap water to remove soil and debris, blotting dry on paper towels, and slicing with a razor blade. Remove the seed tissue from the dehusked cob with a razor blade, and slice. Blot the cut surfaces with Kimwipes for 2–3 sec before printing on the membrane.
3. Lay a nitrocellulose membrane on a piece of 3-mm Whatman paper on a 1-cm stack of paper towels. Press the tissues evenly and firmly onto the membrane for 15–30 sec. Air dry the membrane before staining for total protein (Hancock & Tsang, 1983) or with antibodies (see the following).
4. Block the membrane with 3% BSA solution 1–2 hr.
5. React the membrane with primary antibody for 1 hr.
6. Wash the membrane three times with TTBS for 10 min each.
7. Incubate the membrane with secondary antibody for 1 hr.
8. Wash the membrane three times with TTBS for 10 min each.
9. Wash the membrane with AP buffer for 5 min.
10. Develop the tissue prints with color reagent.

11. Stop with distilled water. Remove the membrane from the water after 1–2 hr, and allow it to air dry. (Note: Leaving the membrane in water overnight will cause excessive coloration of the blot background.)

IV. Localization of Glycine-Rich Protein 1.8 in Bean Ovaries and Hypocotyls[3]

By using tissue printing according to the procedure reported by Cassab and Varner (1987), bean cell wall GRP 1.8 was localized in vascular tissue of hypocotyls and ovaries (Keller & Lamb, 1989). Although there was background staining in the tissue print, the immunoreaction was clearly distinguishable. Besides localizing GRP 1.8 in different organs, tissue printing is also a sensitive and efficient technique for observing GRPs during development: Because a positive immunoreaction in tissue printing depends on a soluble antigen, it has been possible to demonstrate that GRP 1.8 becomes insolubilized in the vascular tissue of bean hypocotyls (Fig. 3.5) 9–11 days after germination (Keller, Templeton, & Lamb, 1989).

V. Tissue Printing to Screen for Mutants: A Search for Cell Surface Carbohydrate Mutants in *Arabidopsis*[4]

Although tissue printing has been mainly used for localizing specific protein and nucleic acid molecules, we have recently used multiple tissue prints of *Arabidopsis* floral stems as a fast and efficient screen for mutants. The method depends on having an appropriate probe

[3]This section was contributed by Beat Keller.

[4]This section was contributed by Nicola Stacey, J. Paul Knox, and Keith Roberts.

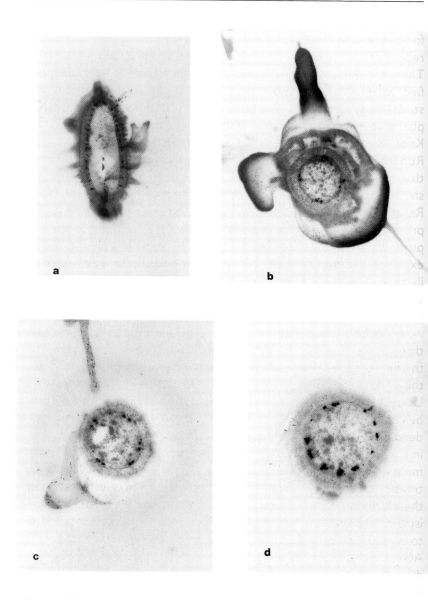

Figure 3.5 Nitrocellulose tissue prints show immunolocalization of GRP 1.8 in cross sections of (a) bean ovary and dark-grown bean hypocotyl at (b) 7 days, (c) 8 days, and (d) 9 days after germination. Prints were reacted with antiserum against GRP 1.8 diluted 1:3000. Bound antibodies were detected after treatment with AP-conjugated anti-rabbit IgG. From "Glycin-Rich Cell Wall Proteins in Bean: Gene Structure and Association of the Protein with the Vascular System" by B. Keller, N. Sauer, and C. J. Lamb (1988). *EMBO Journal* **7**, 3625–3633.

for a molecule test; we have a panel of monoclonal antibodies that recognize a variety of carbohydrate epitopes at the plant cell surface. These include antibodies to methyl-esterified and non-methyl-esterified polygalacturonic acid, JIM5 and JIM7, respectively (Knox, Linstead, King, Cooper, & Roberts, 1990), and antibodies to a set of plasma-membrane-associated AGPs, for example, MAC207 (Pennel, Knox, Scofield, Selvedran, & Roberts, 1989) and JIM4 (Knox, Day, & Roberts, 1989). Many of these antibodies reveal on tissue sections that their corresponding epitopes are developmentally regulated and show precise spatial deposition within particular tissues or cell types. Reacting a simple tissue print with the primary antibody and an appropriate enzyme-linked secondary antibody allows two classes of potential variation to be identified. The first class is over- or underexpression of the particular epitope; such putative mutants are readily seen. The second class includes mutants in which the epitope is still expressed, but in an inappropriate position; these can also be readily identified.

We have found it advantageous to tissue print the same cut face three times (Fig. 3.6). In *Arabidopsis* flowering stems this means that the first print is green with released chlorophyll and the second and third prints contain progressively lower levels of transferred antigens. Using three prints permits subtle variations in the level of antigens to be readily detected, particularly in the last print. It is important to understand that the level of a particular antigen may not reflect its level in the stem but may simply reflect a variation in the degree of its immobilization within the original tissue, thus affecting its availability for transfer to the membrane. Experiments of the kind just described, therefore, should be backed up by conventional immunocytochemistry on the sectioned material. We have used the screening protocol to identify mutants that have altered levels of expression of pectin, AGPs, and a variety of extensin-like proteins. One person can produce triplicate tissue prints at a rate of about 160 plants an hour, whereas with the full protocol it is unlikely that one person could manage more than 300 plants a day with any single probe. Theoretically, the same protocol could be used for screening for a wide variety of mutant classes; for example, substrates for enzymes or labeled lectins of various sorts could be used directly on tissue prints. We have not had any problems transferring any of the carbohydrate-based antigens to the membrane.

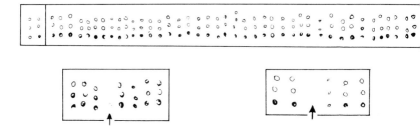

Figure 3.6 The top is a standard mutant screening sheet. Two wild-type *Arabidopsis* floral stems were tissue printed at one end, and triplicate prints of 41 mutagenized plants were printed at the other. The sheet was then reacted with JIM13, a monoclonal antibody to a plasma membrane AGP. The high expression of most of the epitopes is localized in the vascular ring and not at the surface of the *Arabidopsis* stem. The bottom shows portions of two similar screening sheets. The sheet on the left is reacted with JIM19, a monoclonal antibody to an oligosaccharide on an extensin-related protein (courtesy of Neil Donovan). A prospective mutant, showing low-level expression, is indicated by the arrow. The sheet on the right is reacted with JIM13; the arrow indicates the position of a plant showing much reduced expression of the corresponding AGP. Both pictures are simple photocopies of the original sheets; photocopying is an inexpensive and convenient way of keeping records.

A. Materials

1. M2 *Arabidopsis* seed (EMS mutagenized material)
2. Nitrocellulose sheet (Schleicher & Schuell type BA-85) soaked in 0.2 M CaCl$_2$ and blotted dry (use a 1-cm-wide strip for each antibody to be tested)
3. Phosphate-buffered saline (PBS)
4. 10% sheep serum
5. Blocking buffer: 10% sheep serum in PBS
6. Sodium azide, 1% and 0.1%
7. Primary antibody: rat hybridous culture supernatant, diluted 1:20 in PBS containing 10% sheep serum and 0.1% sodium azide
8. Secondary antibody: anti-rat IgG linked to horseradish peroxidase, diluted 1:2000 in PBS containing 10% sheep serum (no azide)

9. Peroxidase substrate: 25 ml water, 5 ml methanol containing 3 g/ml of 4-chloro-1-naphthol, and 30 μl 6% H_2O_2 (make up immediately before use)

B. Procedure

1. Sow *Arabidopsis* seed, one seed per well, in 60-well seed trays. Water compost before sowing seed, which should be suspended in water and transferred individually by pipette. Cover the seed tray with a domed plastic seed tray cover, and do not water again until after germination.

2. Grow plants in a controlled-environment room at 22°C with a 16-hr day length and 50% relative humidity. After 6 wk, floral stems may be used for tissue printing.

3. Cut the floral stem from the plant between the first and second nodes. Press the stem firmly onto the nitrocellulose for 2 sec, three times per nitrocellulose strip. Make a fresh face cut before moving on to next strip if you are making multiple strips for probing with different antibodies.

4. Block all protein binding sites on the nitrocellulose by incubating with 10% sheep serum in PBS. Add 1% sodium azide to inhibit endogenous peroxidases, and incubate in blocking buffer for 1 hr. Perform all incubations at room temperature with agitation.

5. Replace the blocking buffer with the primary antibody. Incubate for 2 hr.

6. Wash three times in PBS, 10 min each wash.

7. Incubate for 1 hr with the secondary antibody for 1 hr.

8. Wash three times in PBS, 10 min each wash.

9. Add the peroxidase substrate to produce a blue color at the binding site. Color should develop within 5 min. Wash thoroughly with water.

10. Dry and examine under a binocular microscope. The first print usually contains a lot of chlorophyll, but the second (or third for abundant antigens) should clearly reveal any differences in the amounts or positions of antigens.

CHAPTER

Induction, Accumulation, Localization, and Transport Studies by Tissue Printing

Elena del Campillo

Department of Plant Biology
University of California at Berkeley
Berkeley, California

I. Overview
II. Immunolocalization of 9.5 Cellulase in Bean Abscission Zones
III. Immunocytolocalization of Polygalacturonase in Ripening Tomato Fruit Using a Membrane That Covalently Binds Protein
IV. Insolubilization of Cell Wall Proteins in Developing Soybean Stems
V. Translocation of a Fungal Protein in *Nicotiana tabacum*

I. Overview

Chemical tissue printing on nitrocellulose membranes, followed by immunochemical staining, has been useful for studying time-dependent

Additional contributions to this chapter have been made by Denise M. Tieman and Avtar K. Handa, Zheng-Hua Ye, Yan-Ru Song, and Joseph E. Varner, Rosannah Taylor, Brian A. Bailey, Jeffrey F. D. Dean, and James D. Anderson.

and tissue-specific accumulations of cell wall proteins in plants. This technique has also been used to study the transport of a protein injected into a plant vascular system.

Chemical tissue prints result from the molecules that transfer from the cut cells of a tissue section to the surface of a synthetic membrane, where they are retained and immobilized. Cell wall proteins make excellent chemical tissue prints because they transfer readily, and with little lateral diffusion, from their natural polysaccharide matrix to a matrix, such as nitrocellulose. The principal steps for tissue printing are (a) the release of the protein of interest from the plant tissue, (b) the contact-diffusion transfer of the protein to the recipient membrane, and (c) the retention and binding of the protein into the synthetic matrix. The protein print is a mirror image of the tissue and can be used to detect and localize specific cell wall proteins. Many cell wall proteins are bound to the cell wall matrix by ionic interactions, yet they can be released by solutions of various ionic strengths. Thus, the release and transfer of cell wall proteins during tissue printing requires using a salt solution. This can be accomplished in two ways: (a) by wetting the membrane with a high-salt buffer before printing or (b) by dipping the tissue slice into a high-salt buffer before printing. The type and concentration of salt depend primarily on the nature of the protein to be transferred. For cell wall hydrolases NaCl is commonly used, but for structural cell wall proteins $CaCl_2$ is more effective, perhaps because Ca^{+2} forms complexes with the pectates and facilitates the release of positively charged proteins. If the protein is already insolubilized in the cell wall matrix, it will not transfer when the tissue is printed, irrespective of the type and amount of salt used.

Retention and binding of a cell wall protein into the recipient matrix depend on the chemistry of the membrane and its ability to interact electrostatically and hydrophobically with the protein of interest. Nitrocellulose membranes have a high binding capacity for proteins and should always be tried first for printing a new type of protein molecule. Adding methanol to the transfer buffer increases the capacity and the affinity of nitrocellulose for proteins, presumably by promoting hydrophobic interactions. However, in some instances nitrocellulose membranes do not bind the protein of interest at all or else bind it only weakly. In these cases, one should next try a membrane that can react chemically with the protein and covalently bind it, such as Immunodyne Immunoaffinity membrane. The type of buffer, salt concentration, and pH of the incubation mixture are important for protein binding and must be determined empirically.

Methods for detecting the cell wall proteins bound to the imprinted membranes must be specific and sensitive. Best results are achieved by using antibodies raised against the cell wall protein of interest. An enzyme-conjugated secondary antibody raised against the primary antibody is commonly used to visualize the binding between the primary antibody and the antigen. To reduce background from nonspecific cross reactions, the primary antibody is combined either with 1% (v/v) normal serum from the species in which the secondary antibody was raised or with a low concentration of SDS or Tween-20. These procedures allow clearer localization of a protein on the tissue print.

Four chemical tissue print protocols that have been successfully used for studying particular cell wall proteins follow. Each was developed by following the principles just discussed.

II. Immunolocalization of 9.5 Cellulase in Bean Abscission Zones

Leaf abscission is primarily a catabolic cell wall process that causes a plant to shed leaves. The petiole fractures after the cell wall dissolves in a narrow layer of cells called the *separation layer*. During leaf abscission in bean plants, a cellulase (endo-β-1,4-glucanase) is induced, and it accumulates to high levels as abscission proceeds. The enzyme has a molecular weight of 51 kDa and a basic pl of 9.5; hence, it is frequently referred to as 9.5 cellulase. This enzyme was purified and used to prepare 9.5 cellulase-specific antiserum. Although the 9.5 cellulase transfers to dry nitrocellulose during tissue printing, moistening the membrane with a solution of methanol and salt produces a clearer localization of 9.5 cellulase on the tissue print. Both dry and moistened nitrocellulose tissue prints immunoblotted with 9.5 cellulase antibody have been used to detect 9.5 cellulase in bean abscission zones (Reid, del Campillo, & Lewis, 1990) and to demonstrate the time-dependent distribution and tissue localization of this enzyme as abcission proceeds (del Campillo, Reid, Sexton, & Lewis, 1990). These studies show that 9.5 cellulase occurs in the cortical cells of the separation layer (Fig. 4.1). It also shows that 9.5 cellulase is associated with the vascular traces of the tissue adjacent to the separation layer (Fig. 4.2).

A. Materials

1. Nitrocellulose membranes, 0.45-μm pore size (Bethesda Research Laboratory), cut into 2×5 cm pieces
2. Glassine paper cut into 2×5 cm pieces
3. 10 mM Tris-HCl, 0.5 M NaCl, and 10% methanol (pH 8.0)
4. 3% gelatin
5. Tween-20
6. Tris buffer saline (TBS): 20 mM Tris-HCl (pH 7.5) and 0.5 M NaCl
7. Tween Tris buffer saline (TTBS): TBS containing 0.05% Tween-20
8. Primary antibody: rabbit anti-9.5-cellulase serum purified on a diethylaminoethyl (DEAE) Affi-gel blue agarose bead column; the enriched immunoglobulin IgG fraction is eluted with a buffer containing 20 mM Tris-HCl (pH 8.0), 28 mM NaCl, and 0.02% sodium azide, and the eluted solution is diluted 1:500 in 0.5% (w/v) gelatin and 1% goat normal serum in TTBS
9. Secondary antibody: goat anti-rabbit IgG (Fc) conjugated with alkaline phosphatase (AP) (Promega) (this antibody reacts only with heavy chains of rabbit IgG)
10. AP substrates: nitro blue tetrazolium (NBT) (Promega) dissolved in 90% dimethylformamide (DMF); 5-bromo-4-chloro-3-indolyl phosphate (BCIP) (Promega) dissolved in DMF
11. AP buffer (APB): 100 mM Tris-HCl (pH 9.5), 100 mM NaCl, and 5 mM MgCl$_2$
12. 5 mM EDTA in 20 mM Tris-HCl (pH 7.5)
13. Bean explants exposed to 25 ppm ethylene to accelerate abscission, harvested at different times during abscission

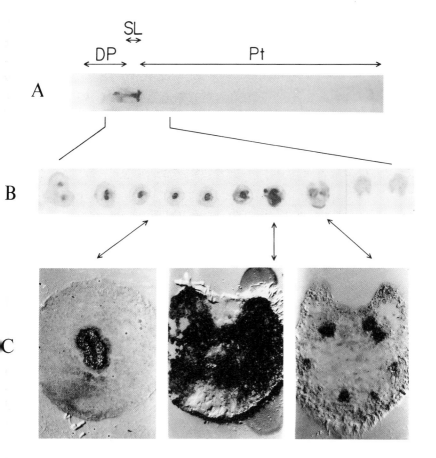

Figure 4.1 Localization of 9.5 cellulase in the distal abscission zone of a bean petiole by tissue printing: (A) longitudinal section of a bean petiole (Pt), including the distal pulvinus (DP) and the distal separation layer (SL); (B) serial cross sections made basipetally through the distal abscission zone, including the pulvinus and 3 mm of the petiole; (C) higher magnification of petiole cross sections, including the separation layer and the tissue immediately above and below. The longitudinal section shows strong staining, primarily in a narrow band of cells perpendicular to the longitudinal axis of the tissue, at the junction of the distal pulvinus and petiole. The serial cross sections show that 9.5 cellulase accumulates predominantly in the separation layer but also appears in the cells of the vascular system of tissues above and below the separation layer.

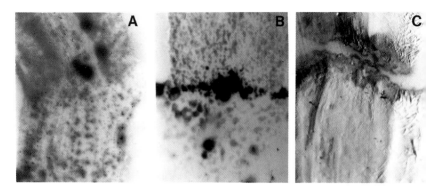

Figure 4.2 Tissue prints of longitudinal sections of bean explants as abscission proceeds. (A) Before abscission develops, tissue prints have no immunostaining for 9.5 cellulase. (B) As abscission develops, 9.5 cellulase appears in the cortical cells of the separation layer. (C) After fracture, 9.5 cellulase appears at both sides of the fracture line.

B. Procedure

1. Wet the nitrocellulose in the 10 mM Tris-HCl, 0.5 M NaCl, and 10% methanol for 5 sec. Place the membrane on a piece of Parafilm on top of a glass plate, and let it dry for 5 min. Then place the nitrocellulose membrane on four layers of Kimwipes.

2. Cut ≈1 mm-thick tissue sections by hand with a sharp razor blade. Make basipetal cross sections through the abscission zone (pulvinus and 5 mm of petiole), making each successive cut with a fresh edge; i.e., cut the top section at the left end of the razor blade, and then progressively shift the blade to the right, so that one razor blade can be used for six or seven sections.

3. Arrange the sections successively over the upper half of the nitrocellulose with a fine tweezer or, alternatively, by simultaneously pushing all the sections off the razor blade with a straightedge.

4. Cover the nitrocellulose with the glassine paper, and press each tissue section with your fingertip as if to make a fingerprint over the glassine paper.

5. Remove the glassine paper (usually the tissue sections come off with it).
6. Arrange the same set of sections in the lower half of the nitrocellulose, cover with a new piece of glassine paper, and press again as indicated in step 4.
7. Rinse the nitrocellulose containing the tissue imprints in TBS for 10 min.
8. Immerse in blocking solution of 3% gelatin for 30 min with shaking.
9. Transfer the nitrocellulose to a fresh TTBS solution, and rinse two times with gentle agitation (5 min per wash).
10. Transfer the nitrocellulose to the 9.5 cellulase antibody solution. Incubate the blots at room temperature for 2 hr with continuous gentle agitation.
11. Wash the blots with TTBS two times, 5 min per wash.
12. Incubate in second antibody, 1:20,000 goat anti-rabbit IgG, for 2 hr at room temperature.
13. Wash the blots two times in TTBS for 15 min at room temperature and then two times in APB for 15 min at 37°C.
14. Add 75 ml of APB containing 25 mg of dissolved NBT and 12.5 mg of dissolved BCIP.
15. When sufficient color has developed ($\approx$5 min), stop the reaction with 5 mM EDTA in 20 mM Tris-HCl.

II. Immunocytolocalization of Polygalacturonase in Ripening Tomato Fruit Using a Membrane That Covalently Binds Protein[1]

Tomato fruit is composed of several tissue types, including those of the pericarp, columella, radial walls, placenta, and seed-containing locular material. These tissues undergo biochemical reactions that lead to

[1]This section was contributed by Denise M. Tieman and Avtar K. Handa.

RIPENING STAGE	TOTAL PROTEIN	PG PROTEIN	ETHYLENE nl/g/h
			0.15
			0.99
			6.85
			9.83
			14.54
			11.88
			6.76

color, flavor, and texture changes during fruit ripening. Polygalacturonase (PG) is a major cell-wall-associated protein induced in ripening tomato fruit, and it may also be a factor in fruit softening. Initial attempts to localize PG by tissue printing on nitrocellulose membranes with polyclonal PG antibodies were not successful. Polygalacturonase, a basic protein with a pI of ≈8.2, did not remain bound to nitrocellulose membranes under normal experimental conditions (Marlow & Handa, 1987). This technical difficulty was overcome by using an Immunodyne Immunoaffinity membrane, which covalently binds protein (Tieman & Handa, 1989). Although higher binding of antibodies to membrane-bound PG was detected in the absence of detergents, SDS was added in the incubation mixture to reduce nonspecific interactions between basic proteins and immunoglobulins (Dimitriadis, 1979). Results indicate that PG first appears in the columella region and then sequentially in the exocarp and endocarp (Fig. 4.3). It is not present in the seed-containing locular material.

A. Materials

1. Immunodyne Immunoaffinity membrane (Pall)
2. Nitrocellulose membranes, 0.45-μm pore size (Schleicher & Schuell type BA-85)
3. Blocking buffer: 0.5% (w/v) nonfat dry milk, 130 mM NaCl, 10 mM sodium phosphate buffer (pH 7.4), 0.001% (v/v) antifoam-A (Sigma), and 0.01% (v/v) thimerosal (Blotto, Sigma)
4. Gelatin solution: 0.25% (w/v) gelatin (Difco), 150 mM NaCl, 5 mM EDTA, 0.05% (v/v), Nonidet P-40 (Sigma), and 50 mM Tris-HCl (pH 7.4)

Figure 4.3 Tissue printing and immunolocalization of polygalacturonase (PG) protein in ripening tomato fruits was accomplished by using Immunodyne Immunoaffinity membranes. Rates of ethylene production ($nl \cdot g^{-1} \cdot hr^{-1}$) at various stages of fruit ripening were as follows: 0.15, mature green; 0.99, breaker; 6.85, turning; 9.83, late turning; 14.54, ripe; 11.88 and 6.76, overripe. From "Immunocytolocalization of Polygalacturonase in Ripening Tomato Fruit" by D. M. Tieman and A. K. Handa (1989). *Plant Physiology* **90**, 17–20.

5. PG antibodies, affinity purified (Marlow & Handa, 1987) and iodinated with Iodo-beads (Pierce) according to manufacturer's instructions
6. Kodak XAR-5 film
7. Tomato fruits at different stages of ripening

B. Procedure

1. Cut each tomato fruit in half with a sharp knife. Place both halves on 3-mm Whatman filter paper saturated with 1 M NaCl for 5 min to enhance release of PG from cell walls.
2. Briefly blot each tomato half on dry 3-mm Whatman filter paper to remove excess moisture from the cut surfaces; this helps to produce clear tissue prints.
3. Place one of the halves on dry Immunodyne Immunoaffinity membrane for 1 min. Periodically apply gentle pressure to the top of the fruit to facilitate transfer.
4. Air dry the tissue print, then transfer it to the blocking solution and incubate for 15 min at room temperature with shaking.
5. Transfer the tissue print to a heat-sealable plastic bag containing 10^6 cpm of ^{125}I-labeled PG-specific antibodies in 10 ml of gelatin solution containing 0.2% SDS and 0.2% Triton X-100. Incubate at 37°C overnight with shaking.
6. Wash the tissue print with five changes of gelatin solution containing no SDS or Triton X-100, for 15 min each, at 37°C with shaking.
7. Air dry the blot, and cover with plastic wrap. Expose to Kodak XAR-5 film at −80°C to obtain the autoradiogram.
8. Place the other tomato half on dry nitrocellulose membrane for 10 min, periodically applying light pressure to enhance transfer of proteins to the membrane.
9. Stain the nitrocellulose membrane containing the tissue print with 0.1% amido black 10B in 25% isopropanol, 10% acetic acid for 1 min, and destain in 25% isopropanol, 10% acetic acid for 30 min with two changes.

Caution: Immunodyne Immunoaffinity membranes can lose binding capacity. Be careful to keep the membranes dry before use. We normally store them in heat-sealed bags in the presence of a desiccant.

V. Insolubilization of Cell Wall Proteins in Developing Soybean Stems[2]

Hydroxyproline-rich glycoproteins (HRGPs), glycine-rich proteins (GRPs), and proline-rich proteins (PRPs) are three classes of cell wall structural proteins in higher plants (Cassab & Varner, 1988). Some time after their secretion into the wall space they are insolubilized in the cell wall by some unknown cross linking. We have used tissue print immunoblots developed by Cassab and Varner (1987), to follow the insolubilization of HRGPs and GRPs in developing soybean tissues (Fig. 4.4) (Ye & Varner, 1991). When the freshly cut tissue is pressed onto a $CaCl_2$-pretreated nitrocellulose membrane, all soluble and salt-soluble proteins are transferred to the membrane. If the wall proteins are cross linked to each other or with other wall components and become insolubilized, they are not transferred and thus cannot be detected by this method. Therefore, tissue print immunoblots, in conjunction with immunocytochemistry, are a unique way to study developmental regulation, synthesis, secretion, and insolubilization of wall proteins in specific tissues.

A. Materials

1. Nitrocellulose membrane, 0.45-μm pore size (Schleicher & Schuell type BA-85)
2. Blocking buffer: 0.1 M Tris-HCl (pH 8.0), 0.05% sodium azide, 0.25% BSA, 0.25% gelatin, and 0.3% Tween-20

---

[2]This section was contributed by Zheng-Hua Ye, Yan-Ru Song, and Joseph E. Varner.

stained sections   HRGPs   GRPs

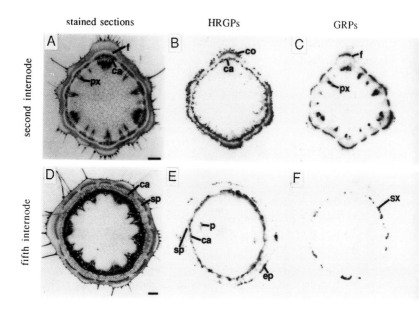

Figure 4.4 HRGPs and GRPs in soybean stem were detected by tissue print immunoblots. Scale bars represent 300 μm. Abbreviations: ca, cambium; co, cortex; primary phloem; ep, epidermis; p, parenchyma; px, primary xylem; sp, secondary phloem; sx, secondary xylem. The deposition and insolubilization of HRGPs and GRPs are developmentally regulated in a tissue-specific manner. In the young soybean stem (A), HRGPs appear mainly in the cambial region and in a few layers of cortex cells surrounding the primary phloem (B), whereas GRPs are localized in the primary xylem and the primary phloem (C). After the stem undergoes secondary growth (D), HRGPs appear mainly in the cambial region, but they are insolubilized in the cortex cells surrounding the primary phloem (E), whereas GRPs appear in the secondary xylem region and are gradually insolubilized in the primary xylem and in the primary phloem (F). The insolubilization of the HRGPs and GRPs in the cell wall was confirmed by immunogold cytochemical localization (Ye & Varner, 1991). The control, which was treated with normal rabbit serum instead of primary antibody, did not show any staining (data not shown). From "Tissue-Specific Expression of Cell Wall Proteins in Developing Soybean Tissues" by Z.-H. Ye and J. E. Varner (1991) *Plant Cell* **3**, 23–37.

3. Rabbit polyclonal antibodies against soybean seed coat HRGP, diluted 1:15,000 in blocking buffer (Cassab & Varner 1987)

4. Rabbit polyclonal antibodies against bean GRP 1.8 fusion protein, diluted 1:15,000 in blocking buffer (Keller, Sauer, & Lamb, 1988)
5. AP-conjugated goat anti-rabbit antibodies (Sigma), diluted 1:20,000 in blocking buffer
6. AP buffer: 0.1 M Tris-HCl (pH 9.5), 0.1 M NaCl, and 5 mM $MgCl_2$
7. AP substrates: 66 μl NBT and 33 μl BCIP (Promega) in 10 ml AP buffer
8. Washing buffer I: 0.1 M Tris-HCl (pH 8.0) and 0.3% Tween-20
9. Washing buffer II: 0.1 M Tris-HCl (pH 8.0), 0.3% Tween-20, and 0.05% SDS
10. Washing buffer III: 10 mM Tris-HCl (pH 8.0) and 1 mM EDTA
11. 0.2 M $CaCl_2$
12. Kodak Technical Pan 2415 film
13. One-month-old soybean plants (*Glycine max* cv. William) grown in the greenhouse

B. Procedure

1. Soak the nitrocellulose membrane in 0.2 M $CaCl_2$ for 30 min, and then air dry before use.
2. Put six layers of No. 1 Whatman paper on a plastic plate. Put one sheet of smooth-surface paper (copier bond is fine) on the Whatman paper, and lay the nitrocellulose membrane on it.
3. Use a double-edged razor blade to cut a 1 mm-thick section of soybean stem. Gently blot the surface of the section on Kimwipes, and then transfer the section onto the nitrocellulose membrane by forceps. Do not move the section after it is transferred onto the membrane. Put four layers of Kimwipes on the section, and press the section gently and evenly for 15–20 sec with your finger. Remove the Kimwipes and the section carefully using forceps. Either keep the section intact or cut a thin section from the tissue for anatomical comparison after staining with toluidine blue.
4. Repeat step 3 for the next prints. Air dry the membrane.

5. Wash the membrane in washing buffer I for 15 min.
6. Incubate the membrane in blocking buffer for 3 hr.
7. Incubate the membrane in the immune sera overnight. Us[normal rabbit serum instead of the immune sera for the neg ative control.
8. Wash the membrane four times with washing buffer I for 1[min each.
9. Incubate the membrane in the diluted AP-conjugated goa[anti-rabbit antibodies for 1 hr.
10. Wash the membrane three times with washing buffer II fo[15 min each.
11. Wash the membrane with washing buffer I for 10 min.
12. Equilibrate the membrane with AP buffer for 10 min.
13. Put the membrane in the AP substrates. Purple color appear[in the place where the protein of interest is localized. Develo[the color until it is satisfactory, generally <10 min. If longe[color development is needed, keep the reaction in the dark.
14. Stop the reaction by washing with buffer III for 5 min. Wash th[membrane in water for 5 min, and then air dry the membrane[

C. Observing and Recording Protein Localization

Observe tissue-level localizations of HRGPs and GRPs on the mem[brane by comparing the physical print on the membrane with th[anatomy of the corresponding section under a dissecting microscope[Record the results on Kodak Technical Pan 2415 film.

V. Translocation of a Fungal Protein in Nicotiana tabacum[3]

Previous studies of pathogen-specific plant proteins showed some o[these proteins to be localized in the plant vascular system (Biles[Martyn, & Wilson, 1989). However, their transport within the vascula[

[3]This section was contributed by Rosannah Taylor, Brian A. Bailey, Jeffrey F. D[Dean, and James D. Anderson.

issues was not demonstrated. Tissue printing on nitrocellulose membranes provides a fast and easy method for screening plant tissues for extracellular proteins produced by pathogens.

Tissue printing on nitrocellulose is used to trace the movement of a fungal protein that elicits plant defense responses in Nicotiana tabacum cv. Xanthi. Specifically, translocation of an ethylene-biosynthesis-inducing endoxylanase (EIX) from the fungus Trichoderma viride was followed from the point of application at a cut petiole. Protein movement was detected in the xylem up and down the stem in sections far removed from the point of application (Fig. 4.5) (Bailey, Dean, & Anderson, 1989, 1991). EIX was detected using polyclonal antibodies raised against a 22 kDa EIX polypeptide (Dean, Gamble, & Anderson, 1989).

Movement of EIX protein through the plant vasculature to the sites of tissue response suggests that the protein itself—not products of its enzymatic activity—is responsible for inducing ethylene biosynthesis and tissue necrosis.

A. Materials

1. Nitrocellulose membrane, 0.45-μm pore size (Schleicher & Schuell type BA-85)
2. Substrate buffer: 1.0 mM MgCl$_2$ in 0.1 M NaHCO$_3$ (pH 9.8)
3. AP substrates: 33 μl BCIP and 66 μl NBT (Promega) in 10 ml substrate buffer
4. Tris buffer: 0.1 mM Tris-HCl (pH 8.0)
5. Washing buffer: 0.1 mM Tris-HCl (pH 8.0) containing 0.5% Tween-20
6. Blocking buffer: 0.1 mM Tris-HCl (pH 8.0) containing 0.25% bovine serum albumin and 0.25% gelatin
7. Goat anti-rabbit AP conjugate (GAR) (Promega), diluted 1:20,000 in blocking buffer
8. Purified EIX protein in distilled water
9. Anti-EIX antibody[4] diluted 1:5000 in blocking buffer

---

[4]Anti-EIX polyclonal antibodies show nonspecific reactions with tobacco proteins. To eliminate these reactions the serum is incubated (1 hr in blocking buffer) with tobacco proteins bound to nitrocellulose membranes (6 × 9 cm). This pretreatment of the antibodies reduces background.

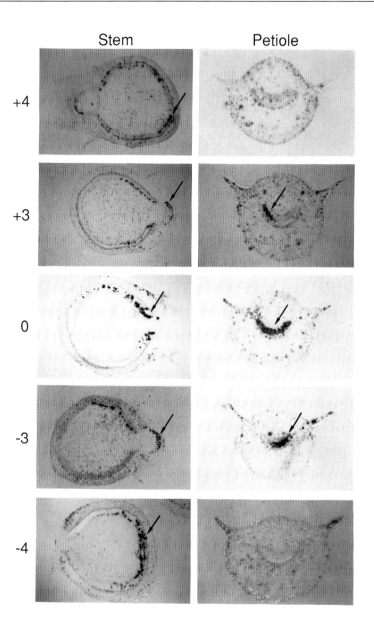

10. *Nicotiana tabacum* cv. Xanthi seedlings cultivated in the greenhouse with natural-day-length illumination until they are 25–30 cm tall and then preincubated in an ethylene atmosphere of 120 µl/liter for 14 hr

Chemicals in the materials list are from Sigma Chemical Company unless otherwise specified.

B. EIX Protein Treatment

After ethylene pretreatment, remove a leaf midway up the plant stem (leaf 0), leaving the exposed petiole. Attach a 3 cm section of Tygon 22 tubing to the exposed petiole, and apply 50 µg of purified EIX protein in distilled water to the petiole through the tubing. Apply boiled EIX to a group of control plants. After 20 min dissect the plant and print as follows.

C. Procedure

1. Make 2–3 mm-thick cross sections of the tobacco petiole and nodal areas of the stem. Preblot each section on a separate piece of membrane or filter paper before printing to remove excess exudate. Place the section on untreated nitrocellulose membrane. Cover with a piece of smooth paper

---

Figure 4.5 Cross sections of tobacco stem nodal tissue and various leaf petioles were printed on nitrocellulose membranes 20 min after EIX application. Negative and positive numbers represent stem nodes and petioles below (−) and above (+) the point of EIX application (0), respectively. EIX was detected with EIX-specific antibodies. EIX was localized within the stem vascular system and was mostly confined to the xylem of the plants. Notice that EIX was restricted to only one side of the plant. EIX proteins were also detected in petiole cross sections of leaves 0, −1, −3, −4, +1, +2, +3, and +4 (not all shown) but were not detected in the petioles of leaves taken from the opposite side of the plant. Leaves displaying symptoms on one side of the leaf blade (leaf +3) generally contained EIX only in the xylem on the same side of the petiole as shown in petiole +3. From "An Ethylene Biosynthesis-Inducing Endoxylanase is Translocated through the Xylem of *Nicotiana tabacum* cv. Xanthi Plants" by B. A. Bailey, J. J. D. Dean, and J. D. Anderson (1991). *Plant Physiology* **97**, 1181–1186.

to protect from fingerprints, and then apply pressure for 10–30 sec.
2. Wash tissue blot for 10–15 min in washing buffer.
3. Replace the washing buffer with blocking buffer, and block 1–2 hr.
4. Replace the blocking buffer with anti-EIX antibody in blocking buffer, and incubate for 3–24 hr.
5. Wash the membrane three times for 15 min each in washing buffer.
6. Wash the membrane once for 15 min in Tris buffer (without the Tween).
7. Incubate the blots in diluted GAR for 1–2 hr.
8. Repeat steps 5 and 6.
9. Develop the blots in AP substrates with mild agitation.
10. Stop the reaction by flooding the blots with distilled water.
11. Dry the blots immediately by slowly blowing warm air over them, and observe.

Acknowledgment

The authors wish to thank H. David Clark for his assistance with the photographic reproductions.

CHAPTER 5

Visualization of Enzyme Activity

Joseph E. Varner

Department of Biology
Washington University
St. Louis, Missouri

I. Overview
II. Enzymatic and Immunological Localization of Mushroom Tyrosinase
III. Ethylene Effect on Peroxidase Distribution in Pea
IV. Tissue Blotting Methods for Identifying Cell-Specific
Enzymes and Antigens

I. Overview

I believe that simple tissue printing techniques are far more capable of qualitatively and quantitatively localizing enzyme activities than any other exploration or exploitation seen so far. My belief is based on the resolution of the locality of protease activity released from mammalian sperm (Gaddum & Blandau, 1970, Fig. 1.2) and on the visualization of the subcellular localization of α-amylase in barley aleurone cells (Jacobsen and Knox, 1973, Fig. 1.3).

Additional contributions to this chapter have been made by William H. Flurkey, Gladys I. Cassab, J.-J. Lin, L.-S. Lin, and Joseph E. Varner, and Daphne J. Osborne.

It is apparent from the tissue prints of Yomo and Taylor (1973) and of Harris and Chrispeels (1975) that protease activity can be easily localized to the nearest cell. Further, it is evident that cytoplasmic protease activity does not significantly smear or diffuse into the wall spaces during sectioning and printing.

Although it has not yet been tried, I believe that tissue printing onto agarose containing the protease substrate might serve well for detecting proteases. In principle, any cellular constituents—macromolecules or metabolites—could be localized on prints on agarose gels containing the components for a suitable assay. The ideal assay would produce or remove an insoluble, colored substance.

II. Enzymatic and Immunological Localization of Mushroom Tyrosinase[1]

Tyrosinase has been localized in many plants, fruits, and vegetables by differential centrifugation, by specific organelle isolation, and ultrastructurally by histochemical and cytochemical methods. On the other hand, tissue-level localization has been more difficult because of endogenous substances that can inactivate enzymes and nonenzymatic oxidation. Localizing mushroom tyrosinase is particularly difficult because of browning reactions that occur at the cut surfaces. Although one can dip mushroom slices into a substrate and look for areas of tyrosinase activity, the results are diffuse and ill defined. Blotting mushroom slices on nitrocellulose is an alternative method for histochemical and immunochemical localization of mushroom tyrosinase (Fig. 5.1) (Moore, Kang, & Flurkey, 1988, 1989; Flurkey & Ingebrigtsen, 1989).

A. Materials

1. Nitrocellulose membranes and 3-mm Whatman filter paper
2. 0.1 M phosphate buffer (pH 6.0)
3. 5 mM L-dopa
4. Catalase

[1]This section was contributed by William H. Flurkey.

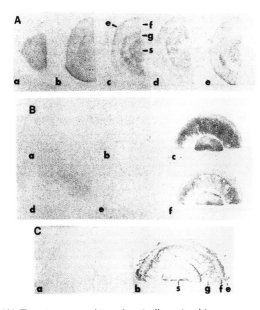

Figure 5.1 (A) Tyrosinase was histochemically stained in cross-sectional slices from a mushroom cap. The blots were incubated in the presence of L-dopa and catalase (100 Keilin units). Slices were taken at intervals from the top of the cap (a) to the bottom (e). In panel c, the letters refer to the cap epidermis or skin (e), cap flesh (f), gill tissue (g), and stalk (s). (B) Tyrosinase was histochemically stained in mushroom blots incubated in various substrate–inhibitor combinations. The blots were incubated in the following solutions: a, 0.5 mM tropolone plus 0.01% H_2O_2; b, 0.5 mM tropolone; c, 5 mM L-dopa; d, 0.5 mM tropolone, 5 mM L-dopa, and 0.01% H_2O_2; e, 0.5 mM tropolone plus 5 mM L-dopa; f, 5 mM L-dopa plus 0.01% H_2O_2. (C) Tyrosinase was immunochemically stained in cross-sectional mushroom tissue slices. A control blot was incubated with mouse antibodies (a) while the blot for localizing tyrosinase was incubated with mouse antibodies and rabbit anti-mushroom tyrosinase (b). The letters refer to the same morphological regions as in A. From "Histochemical and Immunochemical Localization of Tyrosinase in Whole Tissue Sections of Mushrooms" by B. M. Moore, B. Kang, and W. H. Flurkey (1988). *Phytochemistry* **27,** 3735–3737.

5. Blotto (Sigma) and 1–5% goat serum
6. Anti-tyrosinase rabbit serum and goat anti-rabbit (GAR) conjugated with alkaline phosphatase (AP)
7. AP substrates: nitro blue tetrazolium (NBT) and 5-bromo-4-chloro-3-indolyl phosphate (BCIP)
8. Mushrooms

B. Procedure for Enzyme Activity

1. Cut the mushrooms into vertical or cross-sectional slices. Firmly, but lightly and evenly, press the slices onto dry nitrocellulose membrane for 5–20 sec. Several consecutive blots can be made from the same slice without any apparent loss of transfer.

2. Lay the nitrocellulose membrane (blot side up) on 3-mm Whatman filter paper soaked in 0.1 M phosphate buffer containing 5 mM L-dopa and 100 μl catalase. Catalase is included to eliminate peroxidase-catalyzed oxidation of L-dopa. Alternatively, the membrane can be incubated on filter paper soaked with an inhibitor of tyrosinase (such as tropolone or diethyl dithiocarbamate) before incubation with substrate. Allow reactions to proceed for 15 min, during which distinct orange-colored regions should appear. These eventually begin to diffuse together.

3. Remove the membrane from the filter paper and allow to air dry for 30 min. A gray–black stain should appear.

4. Rinse the membrane thoroughly to remove unreacted substrate and soluble products, and then air dry. Use control blots containing buffer instead of substrate solution to monitor nonenzymatic oxidation of L-dopa.

C. Procedure for Immunoblotting

1. Blot mushroom slices on nitrocellulose membrane as previously described.

2. Incubate the membrane in Blotto for several hours, and then incubate in goat serum or goat immunoglobulin IgG fraction for several hours. Goat serum reduces nonspecific background, presumably due to a mushroom lectinlike protein.

3. Incubate the membrane with a primary antibody for several hours.

4. Wash three times and incubate with AP-conjugated GAR serum. AP-conjugated serum is used instead of peroxidase-conjugated secondary antibodies because of endogenous peroxidases in mushrooms.

5. Wash three times, and develop color with AP substrates (NBT and BCIP).

III. Ethylene Effect on Peroxidase Distribution in Pea[2]

Localizing enzyme activity by histochemical methods has been well developed at the microscopic level (Gahan, 1984). Some classical cytochemical methods have been adapted for studying peroxidase activity with tissue prints on nitrocellulose (Spruce, Mayer, & Osborne, 1987; Cassab, Lin, Lin, & Varner, 1988). Various substrates can be used, but only those producing an insoluble end product (3-amino-9-ethylcarbazole, 4-chloro-1-naphthol, 3,3'-diaminobenzidine tetrahydrochloride) are suitable for tissue prints.

The location of peroxidase activity changes during 72 hr of ethylene treatment. Peroxidase activity is found in the vascular bundles in untreated plants; after ethylene treatment, almost no activity is detected in the vascular system, but it is detected in the epidermal and cortical cells (Fig. 5.2).

A. Materials

1. Dark-grown seedlings of *Pisum sativum* cv. Alaska obtained by growing surface-sterilized seeds in vermiculite for 7 days at 25°C and then treating in bell jars with ethylene at a final concentration of 50 ppm
2. Nitrocellulose membranes
3. 0.2 M CaCl$_2$•2H$_2$O
4. 2 mM o-phenylenediamine and 0.012% H$_2$O$_2$ in 60 mM citrate buffer (pH 4.5)

[2]This section was contributed by Gladys I. Cassab, J.-J. Lin, L.-S. Lin, and Joseph E. Varner.

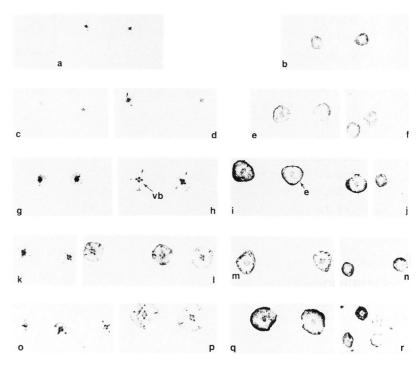

Figure 5.2 Ethylene treatment affects the distribution of extensin and peroxidase in the subapical region of pea epicotyls. Cross sections of pea epicotyls were printed onto nitrocellulose paper. The prints were assayed for peroxidase activity (a, c, d, g, h, k, l, o, p) or reacted with extensin antibody (b, e, f, i, j, m, n, q, r). Ethylene treatment was assessed by examining 7-day-old pea seedlings before treatment (a, b) and ethylene-treated seedlings after 24 hr (d, e), 48 hr (h, i), 72 hr (l, m), and 96 hr (p, q). Untreated pea seedlings 8 days old (c, f), 9 days old (g, j), 10 days old (k, n), and 11 days old (o, r) served as controls. Abbreviations: vb, vascular bundles; e, epidermis. All photographs were taken at the same magnification. From "Ethylene Effect on Extensin and Peroxidase Distribution in the Subapical Region of Pea Epicotyls" by G. I. Cassab, J.-J. Lin, L.-S. Lin, and J. E. Varner (1988). *Plant Physiology* **88**, 522–524.

B. Procedure

1. Soak the nitrocellulose paper in 0.2 M $CaCl_2 \cdot 2H_2O$ for 30 min, and dry on 3-mm Whatman paper.
2. Cut fresh tissue in sections 0.3–3 mm thick with a new razor blade, wash in distilled water for 3 sec, and dry on Kimwipes.

Then, place the section carefully on the nitrocellulose paper with forceps. Press the tissue onto the nitrocellulose paper for 15–30 sec with a gloved fingertip. Finally, carefully remove the tissue section with forceps, and immediately dry the tissue print with warm air.
3. Treat the nitrocellulose paper for the detection of peroxidase activity in 2 mM o-phenylenediamine and 0.012% H_2O_2 in 60 mM citrate buffer.
4. Develop color until a signal appears in the tissue print, and stop the reaction by washing the nitrocellulose paper quickly in distilled water.
5. Dry the nitrocellulose paper on 3-mm Whatman paper.

IV. Tissue Blotting Methods for Identifying Cell-Specific Enzymes and Antigens[3]

Three basic procedures are used in studies of abscission. The simplest and fastest is enzyme-linked immunosorbent assay (ELISA) blotting. For more accurately locating enzymes in specific tissues, histochemical methods may be used after the tissue is blotted on a prepared nitrocellulose membrane. For locating specific antigens to which antibodies are available immunogold procedures may be used after the tissue is blotted on a nitrocellulose membrane.

ELISA BLOTTING

A. Materials

1. 96-well Micro-ELISA plates or 12-well strips (Dynatech)
2. Coating buffer: 1.59 g/liter Na_2CO_3, 2.93 g/liter $NaHCO_3$, and 0.2 g/liter NaN_3 (pH 9.6)

[3]This section was contributed by Daphne J. Osborne.

3. Phosphate buffer saline (PBS): 8 g/liter NaCl, 0.2 g/liter KH$_2$PO$_4$, 2.9 g/liter Na$_2$HPO$_4 \cdot$12H$_2$O, and 0.2 g/liter NaN$_3$ (pH 7.4)
4. Tween PBS (TPBS): PBS plus 0.5 ml/liter Tween-20
5. Blocking buffer: TPBS plus 2% (w/v) polyvinylpyrrolidone (MW 44,000), 0.2% (w/v) egg albumin (both Sigma), 12.5% (w/v) low-fat milk powder
6. Bovine serum albumin (BSA) (Sigma), 5% (w/v) in PBS.
7. Peroxidase immunoconjugate: either GAR/IgG (H+L)/PO or GARa/IgG (H+L)/PO (Nordic), 1:500 in blocking buffer without NaN$_3$
8. Peroxidase substrate: 0.05% (w/v) o-phenylenediamine and 0.03% (v/v) H$_2$O$_2$ in 20 mM sodium acetate (pH 5.0)
9. 3 M H$_2$SO$_4$
10. Mono- or polyclonal antibody with known epitope binding, in BSA solution
11. Micro-ELISA plate reader with 495-nm filter (Dynatech)

B. Procedure

1. Apply 200 μl of coating buffer to each well, and incubate for at least 4 hr at 4°C.
2. For assays of surface washings of intact cells, gently swirl the exposed cell surface of a freshly separated abscission zone in each well for 5 sec, and then remove. Incubate the plate at 4°C overnight.
3. Wash the plate three times with TPBS.
4. Add 200 μl of blocking buffer to each well, and leave for 4 hr.
5. Wash the plate once with TPBS.
6. Probe each well for antigen recognition by adding 200 μl of the antibody solution. Incubate the plate for at least 6 hr at 4°C.
7. Wash the plate three times with TPBS not containing NaN$_3$.
8. Add 200 μl of the appropriate rabbit or rat peroxidase immunoconjugate to each well, and incubate the plate at 4°C for at least 4 hr. Carefully wash the plate three times with TPBS not containing NaN$_3$. Quantify the amount of immunoconjugate bound by adding peroxidase substrate, and either

follow the reaction in time by repeatedly reading the plate recorder or stop the reaction at an appropriate time or color development by adding 50 μl of H_2SO_4.

For assaying antigens at the cut surfaces of sections or in specific pieces of excised tissue, incubate the plate with 200 μl of coating buffer per well overnight at 4°C, empty the plate, and press the sections lightly onto the base of the just-moist wells. Cover the plate (or strip) with plastic film, and incubate another 4–6 hr at 4°C. Wash the plate three times with TPBS, and develop as before.

By using these techniques, the ethylene target cells of abscission zones have been shown to preferentially bind the N-linked monoclonal YZI 2.23 (Fig. 5.3), which recognizes the heptasaccharide epitope Manα3 (Manα6) (Xylβ2) Manβ4 GlcNAcβ4 (Fucα3) GlcNAc (McManus et al., 1988). The polyclonal antibody of pI 9.5 cellulase from *Phaseolus vulgaris* (Koehler et al., 1981) and an avocado mesocarp cellulase (Della-Penna et al., 1986) have also been recognized.

NITROCELLULOSE BLOTTING FOR ENZYME LOCATION

Standard histochemical techniques can be used for this purpose; the best are those that liberate an insoluble and colored or fluorescent product (Gahan, 1984). When the developed tissue blots are viewed under a microscope, the individual cells where the enzyme reactions occur can be located. By using the techniques described here, β-glucosidase activity (and with other substrates, peroxidase activity) has been selectively located within the abscission region of *Elaeis guineensis* fruit (Spruce, Mayer, & Osborne, 1987).

A. Materials

1. Nitrocellulose membrane, 0.45-μm pore size (Schleicher & Schull type BA-28)
2. Substrate: 2-naphthyl-β-D-glucopyranoside (Sigma), 0.03% (w/v) in 20 mM sodium acetate (pH 5.0)

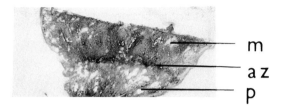

Figure 5.3 Longitudinal section of part of oil palm fruit shows abscission zone cells. The blot was probed with the monoclonal antibody YZI 2.23 to the heptasaccharide epitope Manα3 (Manβ6) (Xylβ2) Manβ4 GlcNAcβ4 (Fucα3) GlcNAc stained with immunogold–silver enhancement. Note the preferential staining of abscission zones. Abbreviations: m, mesocot; az, abscission zone; p, pedicel.

3. Fast Blue B (Sigma), I mg/ml in distilled water
4. Fruit of *Elaeis guineensis*

B. Procedure

1. Soak squares of nitrocellulose membrane (2 × 2 cm) in distilled water, place on microscope slides, and blot surfaces dry with Kimwipes.
2. Place a longitudinal section of the base of the fruit, including the abscission zone, on each membrane, and press lightly into the surface to ensure uniform contact.
3. Carefully remove the section, and wash the membrane surface with distilled water to clear away adhering cell material not immobilized on the membrane and soluble nonproteinaceous compounds from the tissue. Place the section on a dry filter to remove excess water.
4. Transfer each membrane, blot side up, to a glass-fiber disk (Whatman GF/A 2.1 cm) soaked in the substrate (≈200 μl), and incubate in a Parafilm-sealed petri dish for 2–4 hr at 25°C.
5. Remove the substrate by rinsing the membrane in distilled water, and apply Fast Blue B to render the enzyme product visible in 4–5 min.
6. After rinsing again, remove excess water from the membrane with a tissue. Air dry the membrane, and store it between

glass slides in darkness for subsequent microscopic examination or photography.

NITROCELLULOSE BLOTTING FOR ANTIGEN LOCATION

This method allows immunological detection of antigens within specific cells and tissues. For enzyme location the investigator must have the specific monoclonal or polyclonal antibodies to use in conjunction with the method for detecting enzyme activity just described. The Janssen Auroprobe BL with silver enhancement for light microscopy can be used to view the tissue imprints on the nitrocellulose membranes (Fig. 5.3).

A. Materials

1. Nitrocellulose membrane
2. Monoclonal or polyclonal antibody IgGs for detecting specific antigens or enzyme isoforms
3. Auroprobe LM GARa (or GAR) kit (Janssen)
4. IntenSE kit (Janssen)
5. Tris buffer (TB): 20 mM Tris-HCl (pH 7.4), 0.9% (w/v) NaCl, and 0.05% (w/v) Tween-20
6. TB/BSA: 1% (w/v) BSA (Sigma) in TB containing 0.02% (w/v) NaN$_3$
7. Blocking buffer: TPBS plus 2% (w/v) polyvinylpyrrolidone (MW 44,000) (Sigma); 0.2% (w/v) egg albumin (Sigma); 12.5% (w/v) low-fat milk powder; and 1% horse, 1% chicken, and 1% guinea pig serum (v/v) (Sigma)

B. Procedure

1. Blot the membrane as in the previous protocol, and immediately transfer it to TB/BSA.
2. Wash the membrane thoroughly in four changes of solution, and cover with blocking buffer for 2 hr.
3. Rinse with TB.

4. Apply the primary antibody in TB/BSA, and incubate overnight at 4°C in a Parafilm-sealed petri dish on a gentle rocker.
5. Wash both sides of the membrane thoroughly with TB/BSA (this is critical).
6. Apply the IGS LM of the Janssen kit at a dilution of 1:40, and proceed according to the Janssen instructions.

For the silver enhancement stage make sure that the solutions are freshly made up and that the reaction is carried out in relatively low light. It is also preferable to carry out this part of the process at a relatively low temperature (12–15°C on a cooled stage). Stop the reaction immediately when sufficient silver enhancement has taken place.

With this technique McManus and Osborne have demonstrated the preferential distribution of the epitope to YZI 2.23 in abscission zone cells of a number of species (1991), as well as pI 9.5 cellulase in *Phaseolus vulgaris* and the abscission-zone-specific 34-kDa peptide isolated from *Sambucus nigra* (1990).

Lectin and Glycan Recognition

Rafael F. Pont-Lezica

Centre de Physiologie
Université Paul Sabatier
Toulouse, France

I. Overview
II. Glycoconjugate Staining
III. Localization of Specific Sugars
IV. Detection of Lectin Activity with Neoglycoenzymes
V. Detection of Lectin Activity with Fluorescent Sugars
VI. Detection of Potato Lectin with Antibodies
VII. Immunolocalization of Carrageenan Components in Seaweeds

I. Overview

Glycolipids, glycoproteins, and soluble polysaccharides are transferred from cut cells to a membrane when the surface of a freshly cut organ is blotted onto the surface of the membrane. Glycolipids and glyco-proteins are retained by the membrane through hydrophobic interac-tions. However, it is not clear whether the hydrophilic polysaccharides are retained by hydrophobic membranes or are washed out by the

Additional contributions to this chapter have been made by Valerie Vreeland, Mirasol Magbanua, Fraulein Cabanag, Ewinia S. Duran, and Hilconida Calumpong.

71

subsequent treatments. On the other hand, sulfated polysaccharides such as carrageenans, bind to positively charged nylon membranes Nevertheless, several detection methods are available for glycoconjugates, and several protocols that can be applied to tissue prints have been developed.

Unspecific glycoconjugate staining can be achieved directly on the membranes by modifying the periodic acid–Schiff (PAS) reaction that is widely used to detect glycoproteins. In the modification a fluorescent hydrazine replaces the Schiff's (fuchsin-sulfite) reagent and improves sensitivity. The procedure works well and, because glycoconjugates are present in almost every cell, effectively results in an anatomical imprint of the tissue.

Detecting specific sugars in glycoconjugates is far more interesting because it indicates differences in the carbohydrate moiety of glycoconjugates from different cells and tissues. Lectins are probably the most widely used molecules for localizing specific sugars, but antibodies raised against them are also used. Both general and specific information about lectins and their use as molecular probes can be found in several review articles (Roth, 1978; Lis & Sharon, 1986; Liener, Sharon, & Goldstein, 1986; Vasta & Pont-Lezica, 1990). Localization of lectin binding sites has been investigated with light microscopy using fluorescent lectins (Stoddart & Price, 1977; Surek & Sengbusch, 1981; Sengbusch, Mix, Wachholz, & Manshard, 1982) and with electron microscopy using gold-labeled lectins (Roth, 1984; Pavelka & Ellinger, 1985; Benhamou & Oullette, 1986; Hayashi & Ueda, 1987). On the other hand, specific sugars on glycoproteins blotted to a solid matrix after polyacrylamide gel electrophoresis have been detected by using lectin conjugates (Glass, Briggs, & Hnilica, 1981). Affinity-purified lectins are commercially available, as are a variety of lectin conjugates with fluorescent dyes, enzymes, and colloidal gold. The protocol presented in this chapter is similar to those using antibodies as molecular probes: In this case the lectins are themselves labeled, and no secondary antibody is needed to visualize the sugar–lectin complex. Nevertheless, using antibodies raised against a particular lectin is another possible way to detect the sugar–lectin complex.

Lectins are carbohydrate-binding proteins found in almost all organisms, from bacteria to mammals. Localizing particular lectins in specific tissues and following their changes during development are important for understanding the function of lectins. There are two approaches for detecting lectins: One takes advantage of their ability to recognize specific carbohydrates, and the other uses antibodies against the lectin of interest. Both approaches require some basic knowledge about the target lectin, such as its carbohydrate specificity. It is also necessary to have sufficient amounts of purified lectin to raise antibodies. To detect lectin activity by tissue printing, the correct probe must be used, namely, a specific sugar that has the correct anomeric configuration and is linked to a chromophoric group. There are three possibilities:

1. Use fluorescent derivatives of oligosaccharides. Section V describes this method. It should be pointed out that derivatization opens the ring of the sugar at the reducing end, leaving the remaining part of the oligosaccharide in the correct configuration for recognition by the lectin (Fig. 6.4).

2. Use a particular glycoprotein that has an oligosaccharide moiety that can be recognized by the lectin. This method has been used for purifying several lectins by affinity chromatography. The glycoprotein should be labeled; procedures for labeling with fluorescein isothiocyanate (FITC) (Goding, 1983) and for conjugating with alkaline phosphatase (AP) are well known. A number of synthetic glycoproteins are now commercially available (BioCarb). These have been obtained by coupling p-aminophenyl glycosides to human serum albumin (Stowell & Lee, 1980) by carbodiimide. Conjugated antibodies against human serum albumin can then be used to detect the synthetic glycoprotein.

3. Use neoglycoenzymes. This method is based on the same principle as the use of synthetic glycoproteins, namely, attachment of a p-aminophenyl glycoside to a nonglycosylated enzyme. Its advantage is that after incubating with the appropriate neoglycoenzyme, the complex is visualized by enzyme activity. A protocol that illustrates this approach is presented in Section IV (Gabius, Hellmann, Hellmann, Brinck, & Gabius, 1989).

Detecting particular lectins with specific antibodies is a simple method that uses the principles of Western blotting. A protocol in Section VI illustrates the detection of potato lectin in green tissues.

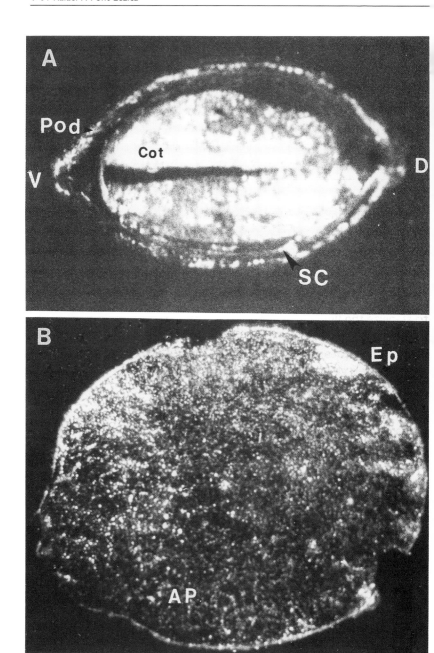

Carrageenans are sulfated galactans abundant in some red algae. The various types of carrageenans differ in their ability to form gels, which is an important commercial property. The protocol in Section VII allows rapid screening of individual plants for the presence of commercially interesting types of carrageenans.

II. Glycoconjugate Staining

Glycoproteins and glycolipids can be stained by a well-known reaction: oxidation of adjacent glycols by periodic acid or its salts to generate two aldehyde groups (Hay, Lewis, & Smith, 1965), followed by the formation of the corresponding hydrazones by reaction with the fluorescent dansyl (DNS) hydrazide (Weber & Hof, 1975; Eckhardt, Hayes, & Goldstein, 1976; Estep & Miller, 1986). A reduction step can be used to stabilize the fluorescent product, but reduction is not essential. The fluorescent derivative may be easily seen under ultraviolet (UV) illumination. This procedure has been used in the detection of soybean and potato tuber glycoproteins by tissue printing (Fig. 6.1) (Pont-Lezica, Taylor, & Varner, 1991).

A. Materials

1. Immobilon P membranes, 0.45-μm (Millipore) (Immobilon P or nylon membranes are preferable to nitrocellulose because they give a lower background)
2. 0.1 M sodium acetate buffer (pH 5.6)
3. 0.5% periodic acid solution

Figure 6.1 (A) In tissue print of soybean pod and seed on Immobilon membrane, reserve glycoproteins in the cotyledons stain strongly. Abbreviations: V, ventral suture; D, dorsal suture; SC, seed coat; Cot, cotyledon. (B) In a potato tuber print, the reserve glycoprotein patatin is uniformly distributed in the amylaceous parenchyma, although the print does not show this uniform distribution because of the difficulty of printing the large organ. Abbreviations: Ep, epidermis; AP, amylaceous parenchyma.

4. DNS hydrazine solution: 0.1 g DNS hydrazine (Sigma) in 33 ml ethanol, diluted to 100 ml with 0.1 M sodium acetate buffer (pH 5.6)
5. 50 mM NaCNBH$_3$ solution (Sigma)
6. Longwave UV light (360 nm)

B. Procedure

1. Wet an Immobilon P membrane with methanol, and transfer it to sodium acetate buffer until ready for use.
2. Put the membrane between several paper towels to absorb excess liquid. The surface of the membrane becomes opaque; at this moment it is ready for printing.
3. Cut a section of tissue about 1 mm thick with a new razor blade, and gently wipe the surface with Kimwipes to absorb excess liquid. Put the freshly cut surface on the membrane, and press for 10–15 seconds. The same tissue surface can be reprinted several times; the successive images will be weaker, but a good imprint can be found among these when glyco-proteins are very abundant in a particular tissue.
4. Wash the membrane twice for 5 min each with gentle agitation to remove unbound material.
5. Incubate the membrane in 0.5% periodic acid solution for 2 hr at room temperature. Make a set of control prints that are not submitted to this step.
6. Wash the membrane with distilled water three times for 1 min each.
7. Incubate the membrane with a freshly prepared DNS hydrazine solution for 2 hr at room temperature.
8. Wash the membrane with 0.1 M sodium acetate buffer several times until the fluorescent background disappears (monitor with the longwave UV light).
9. If the print is photographed immediately, no reduction step is necessary. If not, stabilize the fluorescent derivative by incubating the membrane for 15 min in 50 mM NaCNBH$_3$ solution under a fume hood, and wash with water or sodium acetate buffer. **Caution:** NaCNBH$_3$ is poisonous and flam-

mable. Keep it away from heat, sparks, and open flame, handle it in a hood, and wear gloves and goggles.

10. When the print is dry, observe under UV light with a low-power microscope or magnifying glass. For photography, use ISO 400 film. Black-and-white film requires a yellow filter (Tiffen No. 12), and color film requires a Plexiglas filter to avoid UV irradiation of the film and to maintain color fidelity. **Caution:** Ultraviolet radiation is dangerous, particularly to the eyes. To minimize exposure, make sure that the UV light source is shielded and wear protective goggles or a safety mask to block the UV light.

III. Localization of Specific Sugars

The specificity of lectins toward sugars has been widely used to identify the presence of particular saccharides in glycoconjugates and cell surfaces. For such purposes, a number of affinity-purified lectins are commercially available. These lectins can be obtained as conjugates with (a) fluorescent dyes: fluorescein isothiocyanate (FITC), tetramethylrhodamine isothiocyanate (TRICT), dichlorothriazinyl aminofluorescein (DTAF), or sulforhodamine; (b) enzymes: peroxidase or alcohol dehydrogenase; and (c) colloidal gold. Labeled lectins allow immediate detection of the glycoconjugate–lectin complex by fluorescence, enzyme reaction, or gold–silver enhancement. The following protocol uses commercially available FITC-labeled lectins to localize mannose- and galactose-containing glycoproteins (Fig. 6.2) (Pont-Lezica & Varner, 1989a).

A. Materials

1. Nitrocellulose membranes, 0.45-μm pore size (Schleicher & Schuell). Other membranes may be used.
2. Tween Tris buffer saline (TTBS): 20 mM Tris-HCl (pH 7.5), 0.5 M NaCl, and 0.05% Tween-20.

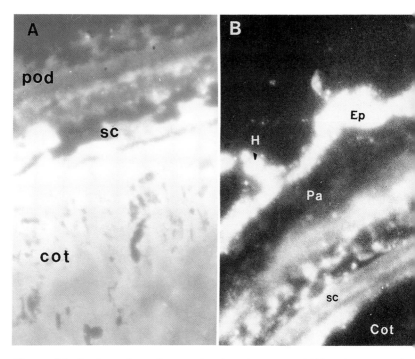

Figure 6.2 Sections of a soybean pod and seed are stained (A) with concanavalin A–FITC to localize α-D-mannose-containing glycoproteins and (B) with *Bandeirae simplicifolia* lectin–FITC to localize α-D-galactose-containing glycoproteins. Cotyledons (Cot) are rich in mannose because soybean reserve proteins have high-mannose *N*-linked oligosaccharides. Galactose-containing glycoproteins are not present in the cotyledons but are abundant in the seed coat (sc), mainly in the pod epidermis (Ep) and hairs (H) (Pa, parenchyma).

3. Blocking buffer: 0.5% periodate-treated bovine serum albumin (BSA) in TTBS. Treat BSA (Sigma, fraction V), 4% in 0.1 *M* sodium acetate (pH 4.5), with 10 m*M* periodic acid for 6 hr at room temperature to inactivate lectin binding sites (fraction V BSA contains glycoprotein impurities). Add glycerol to eliminate excess periodate. Dialyze the solution against TTBS, and dilute to a final concentration of 0.5% BSA.

Alternatively, 3% Tween-20 can be used for blocking to avoid using periodate-treated BSA.

4. Concanavalin A–FITC and *Bandeirae simplicifolia* lectin–FITC conjugates (Sigma), 1 μg/ml in blocking buffer. The concanavalin A–FITC solution contains 10 mM CaCl$_2$.

5. Fluorescent microscope equipped with appropriate FITC filters.

B. Procedure

1. Cut a section of tissue about 1 mm thick with a new razor blade, and gently wipe the surface with Kimwipes to absorb excess liquid. Put the freshly cut surface on the membrane, and press for 10–15 seconds. The same tissue surface can be reprinted several times; successive images will be weaker, but a good imprint can be found among these when the glycoproteins are very abundant in a particular tissue.

2. Transfer the printed membrane to TTBS, and wash the blot two times for 5 min each to remove the unbound material.

3. Block the unoccupied sites of the membrane by shaking the membranes with blocking buffer for 30 min at room temperature.

4. Transfer the membrane to the lectin solution (some lectins require divalent metals for activity, so be sure to add them if necessary). If the lectin is conjugated with a fluorochrome, exclude light from the dish. Incubate at room temperature for 2 hr on a shaker.

5. Wash several times with TTBS (10 min each). Check with a longwave UV light (360 nm) for the disappearance of fluorescent background.

6. To photograph, use a fluorescent microscope with appropriate filters for the fluorochrome employed. If the print is too large for a microscope, use manual UV lights (two lamps to obtain even illumination) and either a yellow filter (Tiffen No. 12) for black-and-white film or a Plexiglas filter for color film to block the UV light. **Caution:** Ultraviolet radiation is dangerous, particularly to the eyes. To minimize exposure, make sure that the UV light source is shielded and wear protective goggles or a safety mask to block the UV light.

IV. Detection of Lectin Activity with Neoglycoenzymes[1]

A neoglycoenzyme is a naturally nonglycosylated enzyme to which a sugar chain has been chemically attached. The chemically glycosylated enzyme combines its specific enzymatic activity with the ability to bind lectins. This property is useful because it allows specific detectors of lectin activity to be synthesized through the correct choice of ligands. Conjugating p-aminophenyl derivatives of carbohydrates to the enzyme by carbodiimide is a simple method that maintains enzyme activity. In this protocol bacterial β-galactosidase is conjugated with p-aminophenyl mannoside to detect lectin activity (Fig. 6.3).

A. Materials

1. β-Galactosidase from E. coli (EC 3.2.1.23) (Boehringer)
2. Phosphate buffer saline (PBS): 20 mM phosphate buffer (pH 7.4) and 0.15 M NaCl
3. p-Aminophenyl-α-D-mannopyranoside (p-AP-Man) and 1-ethyl-3-(3-dimethylaminopropyl)-carbodiimide (EDC), protein-sequencing grade (Sigma)
4. Seeds of Canavalia ensiformis
5. Nitrocellulose membrane presoaked in 0.2 M CaCl$_2$
6. Buffer A: 30 mM Tris-HCl (pH 7.8) containing 150 mM NaCl and 1 mM CaCl$_2$
7. Highly purified BSA (Biomol) treated with periodate to destroy any carbohydrate contaminants, 0.5% in buffer A
8. 0.2 M D-mannose solution
9. β-Galactosidase substrate: Dissolve 5-bromo-4-chloro-3-indolyl-β-D-galactopyranoside (X-Gal, Boehringer) in dimethylformamide, 1 mg/10 μl; add PBS containing 3 mM K$_3$Fe(CN)$_6$ and 3 mM K$_4$Fe(CN)$_6$•3H$_2$O to a final concentration of 1.2 mM X-Gal

---

[1]This section was adapted from "Neoglycoenzymes: A Versatile Tool for Lectin Detection in Solid-Phase Assays and Glycohistochemistry" by S. Gabius, K.-P. Hellmann, T. Hellmann, U. Brinck, and H.-J. Gabius. (1989). *Analytical Biochemistry* **182**, 447–451.

A   B   C

Figure 6.3 Tissue prints from *Canavalia ensiformis* seeds on nitrocellulose are stained for total protein with amido black (A) and for mannose-specific receptors with mannosylated β-galactosidase (4 μg/ml) in the absence (B) or presence (C) of inhibitor, namely, 0.2 *M* D-mannose. Color development was caused by enzymatic activity on X-Gal. From "Neoglycoenzymes: A Versatile Tool for Lectin Detection in Solid-Phase Assays and Glycohistochemistry" by S. Gabius, K.-P. Hellmann, T. Hellman, U. Brinck, and H.-J. Gabius 1989. *Analytical Biochemistry* **182**, 447–451.

B. Preparation of Neoglycoenzyme

1. Dissolve 8 mg *E. coli* β-galactosidase in 1 ml PBS.
2. Add 16 μmol *p*-AP-Man and 32 mmol EDC, and incubate 16 hr at 4°C.
3. Dialyze against PBS at 4°C.
4. Isolate the modified enzyme by gel filtration. Pool the enzyme-containing fractions, dialyze against 60% glycerol in PBS, and store at −20°C.

C. Procedure

1. Keep the *Canavalia ensiformis* seed in a humid atmosphere for 2 days.
2. Presoak a nitrocellulose membrane in a 0.2 *M* CaCl₂ solution for 30 min. Dry the membrane on paper towels at room temperature.
3. Cut the seed with a new razor blade, print the freshly cut surface on the membrane, and press for 10–15 sec.
4. Block the remaining sites by incubating the membrane with 0.5% BSA for 20 min. Wash the print with buffer A six times, 3 min each.

5. Incubate the blot with the mannosylated neoglycoenzyme (
    μg/ml) for 1 hr in the absence or presence of a competitive
    inhibitor (use 0.2 M D-mannose for controls).
6. Wash the membrane several times with buffer A.
7. Visualize the lectin–neoglycoenzyme complex by incubating
    the nitrocellulose with the β-galactosidase substrate.
8. To detect total proteins, stain the blot with amido black.

V. Detection of Lectin Activity with Fluorescent Sugars

Lectins present in different tissues can be detected through their ability to bind specific sugars. In this protocol, DNS hydrazide is used to
synthesize a fluorescent hydrazone of a reducing sugar (Alpenfels
1981; Muramoto, Goto, & Kamiya, 1987). An oligosaccharide must be
used because the formation of the hydrazone opens the lactone ring of
the sugar at the reducing end (Fig. 6.4). The attached mono- or disaccharide at the nonreducing end will conserve the ring structure that is
recognized by the lectins. The procedure has been used for detecting
lectin activity in soybean seeds and potato tubers (Fig. 6.5) (Pont-
Lezica, Taylor, & Varner, 1991).

A. Materials

1. Nitrocellulose membrane, 0.45-μm pore size (Schleicher &
    Schuell) (another membrane may be used)
2. Tween phosphate buffer saline (TPBS): 10 mM phosphate
    buffer (pH 7.4), 0.15 M NaCl, and 0.05% Tween-20
3. Materials for synthesis of fluorescent sugar (N,N',N''-triacetyl
    chitotriose DNS hydrazone or lactosyl DNS hydrazone)
    a. Oligosaccharide: N,N',N''-triacetyl chitotriose or lactose
    b. 40 mM DNS hydrazine in 1% ethanol

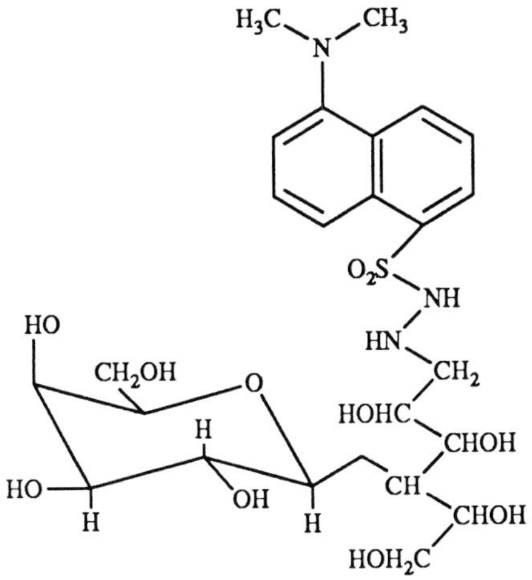

Figure 6.4 In an oligosaccharide, formation of the hydrazone opens the lactone ring of the sugar at the reducing end: (A) lactosyl DNS hydrazone (4-O-β-galactopyranosyl-glucosyl DNS hydrazone) and (B) N,N',N''-triacetyl chitotriose DNS hydrazone [2-acetamido-2-deoxy-4-(2-acetamido-2-deoxy-β-D-glucopyranosyl)$_2$-D-glucosyl DNS hydrazone].

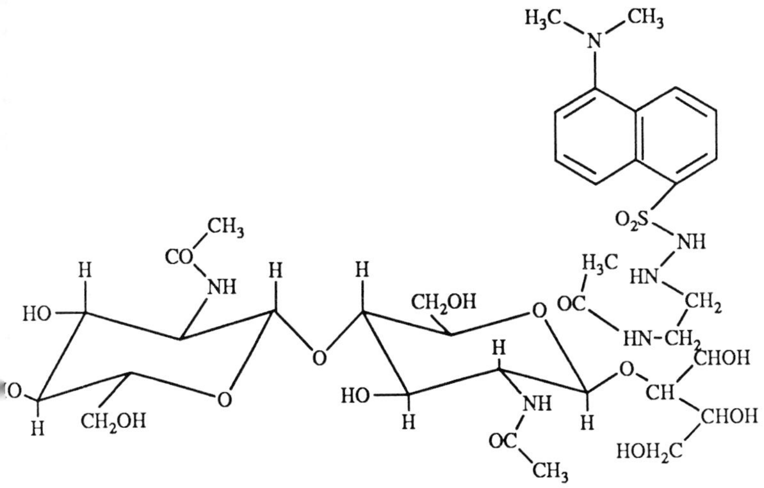

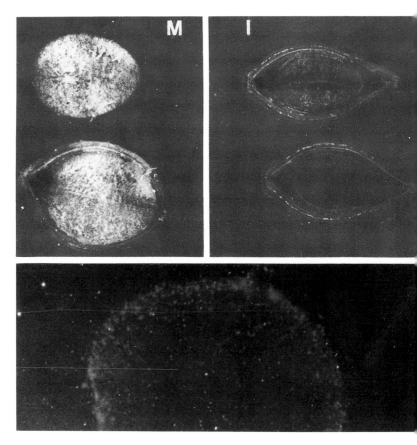

Figure 6.5 Upper panel shows prints of mature (M) and immature (I) soybean poc and seed incubated with lactosyl DNS hydrazone. Binding of fluorescent ligand i observed only in mature seeds. This finding agrees with the known pattern of lecti accumulation in cotyledons at late stages of seed development. The lower panel is print of a potato tuber section incubated with N,N',N''-triacetyl chitotriose DN hydrazone. The fluorescent ligand binds preferentially to the epidermic an subepidermic tissues, where the tuber lectin is more abundant. From "Solanun *tuberosum* Agglutinin Accumulation during Tuber Development" by R. F. Pont-Lezica R. Taylor, and J. E. Varner, (1991). *Journal of Plant Physiology* **137**, 453–458.

    c. 10% trifluoroacetic acid
    d. 50 mM NaCNBH$_3$ solution (Sigma)
    e. SEP-PAK C$_{18}$ cartridges (Waters)
    f. Acetonitrile

B. Synthesis of Fluorescent Sugar

1. Dissolve 10 mg of the oligosaccharide in 200 μl of water.
2. Add 400 μl of 40 mM DNS hydrazine in 1% ethanol plus 40 μl 10% trifluoroacetic acid.
3. Incubate at 60°C for 20 min.
4. Cool the sample, add 30 μmol of NaCNBH3, and incubate 20 min at room temperature in a hood. **Caution:** NaCNBH$_3$ is poisonous and flammable. Keep it away from heat, sparks, and open flame, handle it in a hood, and wear gloves and goggles.
5. Add water to bring the total volume to 4 ml.
6. Activate a SEP-PAK C$_{18}$ cartridge by rinsing with 2 ml of acetonitrile followed by 2 ml of water.
7. Pass the sample through the activated cartridge, and rinse with 2 ml of 10% (v/v) acetonitrile in water.
8. Elute the DNS hydrazone with 2 ml of 40% acetonitrile.
9. Freeze-dry the eluted DNS hydrazone, and resuspend it in PBS (0.1 mM concentration).

C. Procedure

1. Cut a section of tissue about 1 mm thick with a new razor blade, and gently wipe the surface with Kimwipes to absorb excess liquid. Put the freshly cut surface on the membrane, and press for 10–15 seconds. The same tissue surface can be reprinted several times; the successive images will be weaker, but a good imprint can be found among these when the glycoproteins are very abundant in a particular tissue.
2. Transfer the printed membrane to TPBS, and remove the unbound material by washing twice for 5 min each with gentle shaking.

3. Transfer the membrane to the fluorescent sugar solution and incubate 3 hr at room temperature.
4. Wash the print with TPBS several times until the fluorescent background disappears.
5. To photograph, use a fluorescent microscope with appropriate filters for the fluorochrome employed. If the print is too large for the microscope, use manual UV lights (preferably two) and either a yellow filter (Tiffen No. 12) for black-and white film or a Plexiglas filter for color film to block the UV light. **Caution:** Ultraviolet radiation is dangerous, particularly to the eyes. To minimize exposure, make sure that the UV light source is shielded and wear protective goggles or a safety mask to block the UV light.

VI. Detection of Potato Lectin with Antibodies

Lectins can be localized in specific tissues by using polyclonal or monoclonal antibodies. The procedure for obtaining polyclonal antibodies against the native and deglycosylated potato tuber lectin was described by Pont-Lezica, Taylor, and Varner (1991). Characterization of these antibodies indicates that they cross react with related solanaceous lectins, such as tomato lectin. The antibodies against the native lectin recognize the hydroxyproline-rich glycoprotein extensins (27% cross reactivity by competitive ELISA) but not arabinogalactan proteins. The cross reactivity is due to the similar glycosylation pattern of these proteins, because the antibodies against the deglycosylated lectin do not react with extensins. Localization of phloem-specific lectin in potato stems is illustrated in Fig. 6.6.

A. Materials

1. Nitrocellulose membrane, 0.45-μm pore size (Schleicher & Schuell) (another membrane may be used)

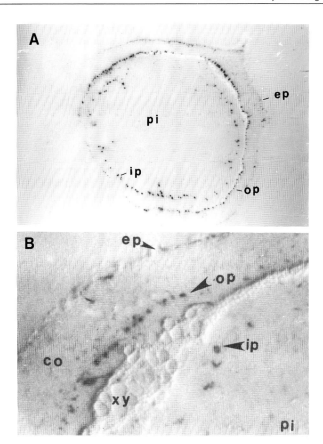

Figure 6.6 Tissue prints of potato stem were incubated with antibodies raised in rabbit against the deglycosylated lectin, followed by goat anti-rabbit AP: (A) transverse section; (B) magnification of a stem sector with lateral illumination showing greater detail of lectin localization. The lectin is present only in the outer phloem (op) and inner phloem (ip); the color at the epidermis (ep) is not a positive reaction but a transfer of natural pigments (xy, xylem; co, cortex; pi, pith parenchyma).

2. TTBS: 20 mM Tris-HCl (pH 7.5), 0.5 M NaCl, and 0.05% Tween-20
3. Blocking buffer: 0.25% gelatin and 0.25% BSA in TTBS.
4. Antibodies against potato lectin, in blocking buffer (1:15,000 dilution for the antibody against the native lectin; 1:5000 dilution for the antibody against deglycosylated lectin)

5. Goat anti-rabbit (GAR) AP conjugate $F(ab)^2$ fragmen (Sigma), 1:20,000 dilution in blocking buffer
6. AP buffer: 0.1 M Tris-HCl (pH 9.5), 0.1 M NaCl, and 5 mΛ $MgCl_2$
7. AP substrates: nitro blue tetrazolium (NBT), 50 mg/ml in 70% methanol diluted 33:10,000 in AP buffer, and 5-bromo 4-chloro-3-indolyl phosphate (BCIP), 50 mg/ml in dimethyl formamide diluted 66:10,000 in AP buffer
8. Potato stem

B. Procedure

1. Cut a section of the tissue about 1 mm thick with a new razor blade, and gently wipe the surface with Kimwipes to absorb excess liquid. Put the freshly cut surface on the membrane, and press for 10–15 seconds. The same tissue surface can be reprinted several times; the successive images will be weaker, but a good imprint can be found among these when the proteins are very abundant in a particular tissue. To avoid interference from extensin-type proteins, do not soak the nitrocellulose in $CaCl_2$, as indicated for other cell wall proteins (Cassab & Varner, 1987).
2. Transfer the printed membrane to TTBS, and wash the unbound material two times for 5 min each in a shaker.
3. Block the unoccupied sites of the membrane by shaking the membrane in blocking buffer for 30 min at room temperature.
4. Transfer the membrane to the first antibody solution (anti-lectin serum), and incubate overnight in a refrigerator or for 2 hr at room temperature. For controls either use pre-immune serum at the same dilution or skip the first antibody step.
5. Wash the membrane in TTBS three times for 10 min each.
6. Incubate the prints with the secondary antibody (GAR–AP) for 2 hr at room temperature.
7. Wash the membrane several times with TTBS (five washes of 5 min each). The last wash is made with TBS without detergent.

8. Incubate the prints in AP substrates at room temperature until the reaction product is observed; treated and control prints should be incubated under the same conditions. Occasionally some tissues show cross reaction with the goat serum. To avoid such a reaction, two procedures are available: (*a*) Block the prints with blocking buffer containing a 1:3000 dilution of normal goat serum or (*b*) use a secondary antibody raised on a different animal. Occasionally a particular tissue contains reducing substances not washed out of the nitrocellulose. These compounds can reduce the tetrazolium salt to the formazan precipitate in the absence of the enzyme. In such cases, using an alternative procedure, e.g., GAR–gold (with silver enhancement) or GAR–FITC, is recommended.

VII. Immunolocalization of Carrageenan Components in Seaweeds[2]

Carrageenans, a family of variable sulfated galactans, are the major cell wall component in some red algae. These linear polymers are based on a D-galactose disaccharide repeat with alternating α-1,3 and β-1,4 linkages. Sulfation sites and anhydrogalactose contents differ for kappa, iota, lambda, and other carrageenan types. Kappa forms firm gels, iota forms soft gels, and lambda is nongelling.

Kappa- and iota-carrageenophytes cultivated in the Philippines are commercial sources of carrageenan for food and industrial applications. The cultivated plants reproduce vegetatively (Azanza-Corrales & Dawes, 1989). They also sporulate to produce wild plants (Doty, 1985, 1986; Azanza-Corrales, 1990), possibly including hybrids that could contaminate commercial stock and decrease the carrageenan gel's strength (Azanza-Corrales, 1991) when they are included with cultivated plants by subsistence farmers.

Tissue printing can screen potential seed stock on reef platforms in 1–2 hr. Carrageenans bind to quaternary amino groups on positively

[2]This section was contributed by Valerie Vreeland, Mirasol Magbanua, Fraulein Cabanag, Erwinia S. Duran, and Hilconida Calumpong.

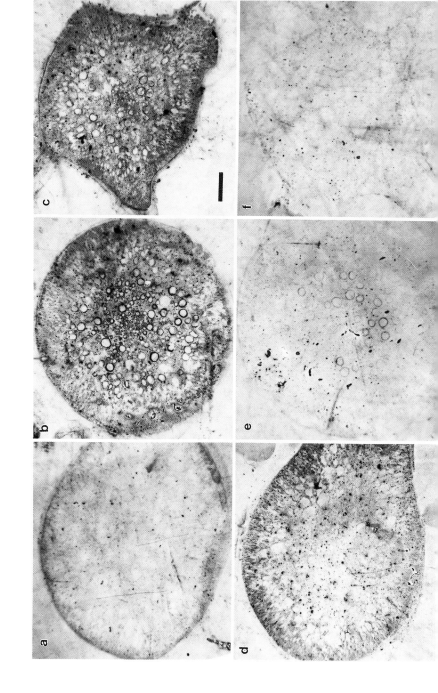

charged nylon membranes for immunolabeling of tissue prints. We prepared monoclonal antibodies as molecular markers for carrageenan types, and immunofluorescence showed differential extracellular localization of carrageenan antigens in *Kappaphycus* (Vreeland, Zablackis, & Laetsch, 1988). The anti-carrageenan monoclonal antibodies are primarily the immunoglobulin IgM, like many anti-carbohydrate antibodies. Second antibodies to the light chain of the monoclonal antibody (a kappa light chain in most cases) provide much stronger labeling than do those to the anti-mu heavy chain (V. Vreeland & X. L. Wang, unpublished data).

Tissue printing of cultivated kappa- and iota-carrageenophytes followed by labeling with monoclonal antibodies to carrageenan clearly differentiates the two kinds of commercial stock. Tissue printing is being used to identify individual plants with pure or mixed carrageenan types (Figs. 6.7 and 6.8).

A. Materials

1. Biotrans B nylon membrane, 0.2-μm pore size (Pall BNAZF3R).
2. 1% ovalbumin (Sigma grade 2) in 50 mM Tris (pH 7.5) and 0.12% sodium azide. (Centrifuge at 12,000 g, and store aliquoted at −20°C.)
3. Wash solution: 200 mM NaCl and 10 mM CaCl$_2$ adjusted to pH 7.0 with NaOH.
4. Anti-carrageenan monoclonal antibodies in hybridoma culture medium (Vreeland, Zablackis, & Laetch, 1988)

Figure 6.7 Tissue prints of cross sections of a brown kappa-producing seaweed (a, d), a brown iota-producing plant (b, e), and a brownish green apparent hybrid seaweed (c, f) are labeled with two anti-carrageenan antibodies, 26-3G1-1C4 to lambda carrageenan (a–c) and 26-6A11-2A4 to kappa carrageenan (d–f). The hybrid plant had the branching morphology of a kappa-producing plant but also had a few short spines on the main axis, similar to the numerous spines on the iota-producing plant. The plants were collected at the FMC seaweed farm at Tindog Beach, Cebu, Republic of the Philippines, and were identified by Rubin Barraca of FMC. Prints were made from fresh plants on the day of harvest; they were air dried and labeled several weeks later in Berkeley, California. The scale bar represents 1 mm.

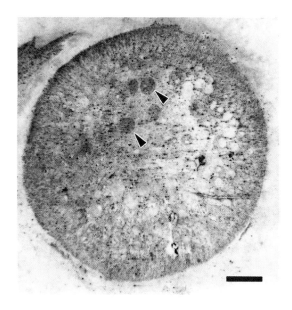

Figure 6.8 Antibody 26-6A11-2A4 to kappa carrageenan labels intracellular carrageenan (arrowheads) inside some large (0.2–1.0-mm diameter) medullary cells in a formaldehyde-fixed, large-diameter green kappa-producing seaweed. The scale bar represents 1 mm. Immunogold ultrastructural localization revealed intracellular synthesis of carrageenan components on embedded sections of a red alga (Gretz, Wu, Vreeland, & Scott, 1990).

(hybridoma medium contains Iscove's modified Dulbecco's Eagle's medium with 20% fetal or newborn bovine serum).

5. Second antibody: AP-conjugated goat anti-mouse antibody, kappa chain specific (Fisher). Dilute 1:1000 in hybridoma medium.

6. BCIP stock solution: 4 mg/ml BCIP *p*-toluidine salt (Sigma) in 1:2 acetone:methanol stored at −20°C. (Note: If a precipitate forms in the stock solution, it can be dissolved by briefly warming. If its activity is low or a dark bluish background forms, the precipitated stock should be discarded.)

7. Substrate buffer: 200 mM Tris and 4 mM $MgCl_2$ (pH 9.5)

(24.2 g Trizma base plus 810 mg $MgCl_2 \cdot H_2O$ diluted to 1:1 with distilled water).

8. Substrate solution: 0.75 ml of BCIP stock solution and 5 mg of NBT, grade 3 (Sigma), added to 50 ml of substrate buffer immediately before use. (The substrate solution can be stored in the dark, preferably in the refrigerator, for reuse on the same day. It should be discarded when a blue or purple precipitate forms.)

9. *Kappaphycus alvarezii* = *Eucheuma alvarezii* Doty (produces kappa carrageenan; commercial name, *E. cottonii*) and *Eucheuma denticulatum* (Burman) Collins et Hervey (produces iota carrageenan; commercial name, *E. spinosum*). Fresh, frozen, dried, or formalin-fixed plants may be used for printing, but fresh plants give the sharpest prints. Dried plants should be washed briefly in tap water and then rehydrated for 30 min in 500 m*M* KCl plus 50 m*M* $CaCl_2$ before printing.

10. Kodak Technical Pan 2415 film (Estar base), ISO 125–160; Kodak HC-110 developer, dilution B

B. Procedure

1. Using fine forceps and vinyl gloves for handling, cut the Biotrans B membrane with scissors to the minimum usable size. Identify the membrane by a pencil notation in a corner, and place it on a piece of Whatman No. 1 filter paper.

2. Cut a slice of tissue 1–2 mm thick on a polyethylene or other stiff plastic sheet with a clean, sharp single-edged razor blade. Transfer the tissue to the membrane with fine forceps, taking care not to move the tissue after first contact with the membrane. Finger-press the tissue firmly for 30–60 sec with a clean glove, and then remove the tissue with forceps. Air dry the print.

3. Block the membrane by incubation for 15 min in ovalbumin with gentle agitation on a rotary shaker. Rinse briefly in wash solution. Blot excess liquid from the edge of the membrane onto a paper towel.

4. Place the damp membrane (print side up) on a dry plastic surface in a small covered container to prevent evaporation.

A six-well tissue culture plate is convenient for multiple antibody incubations. Incubate the membrane in a minimal volume (25–50 µl/cm²) of monoclonal antibody for 15 min. Transfer the hybridoma supernatant with a sterile disposable pipette tip to avoid contaminating the antibody supply.

5. Wash the membrane briefly, and then transfer it to a larger volume of fresh wash solution. Gently agitate it for 4–5 min. Remove excess liquid from the membrane.

6. Cover the damp membrane with diluted second antibody, and incubate 15 min.

7. Wash as in step 5.

8. Incubate several membranes in the substrate solution for 5–30 min with gentle agitation until the background darkens.

9. Stop color development by a brief wash in distilled water. The results should not be analyzed until the print is dry, because the background lightens upon drying and wall impressions appear dark until they are completely dry.

10. Photograph the BCIP/NBT-stained tissue print with Kodak Technical Pan film. Develop the film in Kodak HC-110 developer for 8.5 min at 22–23°C for high contrast.

CHAPTER 7

Gene Expression in Plants

Zheng-Hua Ye*, Yan-Ru Song†, and Joseph E. Varner*

*Department of Biology
Washington University
St. Louis, Missouri

†Institute of Botany
Academia Sinica
Bejing, China

I. Overview

When a freshly cut section of plant tissue is pressed onto a nitrocellulose or nylon membrane, the soluble contents of the cut cells are transferred to the membrane. The cell contents make a latent print on the membrane that can then be visualized with appropriate probes. The rigid cell walls (especially xylem and phloem fibers) leave

Additional contributions to this chapter have been made by Tom J. Guilfoyle, Bruce A. McClure, Melissa A. Gee, and Gretchen Hagen, Mark L. Tucker, R.W.F. Harriman, Denise M. Tieman, and Avtar K. Handa.

a physical print that makes the anatomy of the tissue visible on the membrane without any further treatment. This technique, called *tissue printing*, was first developed by Cassab and Varner (1987) to localize cell wall extensin in soybean seeds. The technique was further developed to localize extensin mRNA in soybean pods and α-subunit mRNA of β-conglycinin in soybean seeds at different developmental stages with ^{32}P-labeled DNA probes (Varner, Song, Lin, & Yuen, 1989) and to study the tissue-specific expression of auxin-regulated genes in elongating hypocotyl regions of etiolated soybean seedlings and the rapid turnover of RNAs encoded by these genes during gravistimulation with ^{35}S-labeled antisense RNA probes (McClure & Guilfoyle, 1989a, 1989b). Although tissue printing has lower resolution than *in situ* hybridization for localization of mRNAs, it is a quick and convenient way to screen a large number of tissues from different developmental stages or from different plants at the same time on the same membrane. It facilitates studying the spatial distribution of mRNAs in large organs, such as tomato fruit (Handa, Harriman, & Tieman, Section VI) and whole seedlings, and it allows tissue-level localization of mRNAs.

Both nitrocellulose and nylon membranes can be used for tissue prints without being pretreated, but nylon membranes are generally preferred because they are easier to handle. The principle for retaining transferred mRNAs on the membrane is the same as that for RNA gel blots and dot blots. Tissue printing, however, has the advantage that freshly cut tissue can be immediately pressed onto the membrane without any treatment. Cellular mRNAs are immobilized on the membrane and are then detected by specific probes. The specific cellular mRNAs show little lateral diffusion during or after printing. The mRNAs can be precisely located by comparing the print with the anatomy of corresponding stained sections or with the physical print on the membrane. Another advantage of tissue printing is that large organs or whole seedlings can be printed on the same membrane, allowing the spatial distribution of specific mRNAs to be analyzed. Song and Varner's protocol (Section III) demonstrates localization of extensin mRNA in the seed coat regions of intact tomato fruit by tissue printing and probing with ^{35}S-labeled extensin antisense RNA. The protocol of Gee, Guilfoyle, Hagen, and McClure (Section II) shows how to determine the distribution of small auxin up-regulated RNAs in gravistimulated soybean seedlings by simply cutting the whole seedling longitudinally, pressing the cut surface on the membrane, and then probing with specific probes.

Tissue printing is extremely useful when the sample is large, e.g., a whole fruit. The protocol of Handa, Harriman, and Tieman (Section VI) uses Northern prints of tomato fruits to localize pectin methylesterase mRNA during ripening. Baird, Sexton, and Tucker (Section V) show how to determine the distribution of cellulase mRNA in bean tissue by printing longitudinal and cross sections of the bean petioles on the membrane and then probing with cellulase antisense RNA. Normally, many prints may be done on the same membrane, so tissue printing is a wonderful tool for screening for organ- and tissue-specific gene expression in different plants and different developmental stages simultaneously. Also, because all the prints are on the same membrane and are treated with the same hybridization, washing, and exposure, the specific mRNAs from different developmental stages can be roughly quantified by comparing the signal intensities of the different prints on the membrane.

For tissue print hybridization, first the section is cut and then the cut surface is pressed onto the membrane. To obtain good prints (a) each razor blade edge should be used only once, because the surface of the blade is contaminated after cutting; (b) the sections should be transferred with forceps; (c) the sections should be pressed onto the membrane carefully and evenly to avoid crushing; and (d) the surfaces of the sections should be blotted gently with Kimwipes if the tissue is juicy or if a copious amount of fluid comes from vascular bundles. Generally, it is advantageous to print several sections so that the signals of the prints can be compared to see whether they show the same pattern. Whether you have obtained an even physical print or whether there is the same distribution of chlorophyll in the print as in the section (if there is chlorophyll) can be conveniently determined by examining the print with a hand lens. The printed membrane may be stored up to several months at $4°C$ without loss of signals. In all procedures involving RNA, one should take precautions to prevent RNase contamination; wear gloves when you make the prints.

Both ^{32}P-labeled DNA probes and ^{35}S-labeled RNA probes have been used for tissue print hybridization (Varner, Song, Lin, & Yuen,

1989; McClure & Guilfoyle, 1989a, 1989b). The choice of probe depends on the tissue size used and the resolution desired. ^{32}P-labeled probes give very high specific activity, making them suitable for use with x-ray film autoradiography in identifying organs or regions of organs containing individual mRNAs. However, their resolution is usually insufficient for precisely localizing specific labeled cells. As for *in situ* hybridization, ^{35}S-labeled probes are preferred for tissue print hybridization because of the high specific activity attainable, their high autoradiographic efficiency, their reasonable half-life, and their safety. Their resolution is usually adequate for tissue-level localization of specific mRNAs; e.g., in a 2-mm print of soybean stem cross section, it is easy to distinguish the glycine-rich protein (GRP) mRNA hybridization signal in primary xylem from the hydroxyproline-rich glycoprotein (HRGP) mRNA hybridization signal in the cambial region, as is shown in Section IV. DNA probes radiolabeled to high specific activity can be prepared by random oligonucleotide primed synthesis (Sambrook, Fritsch, & Maniatis, 1989). RNA probes can be prepared by cloning the corresponding DNA fragment into a vector containing RNA polymerase promoters (e.g., $_p$Bluescript vector from Stratagene) and synthesizing strand-specific RNA by corresponding RNA polymerase (Sambrook, Fritsch, & Maniatis, 1989). For making strand-specific RNA probes, plasmid templates should be linearized with an appropriate restriction endonuclease. Generally, single-strand RNA probes are preferred over DNA probes because the higher thermal stability of RNA–RNA duplexes allows more stringent wash conditions, thus resulting in lower nonspecific hybridization background.

The choice of the film depends on the tissue size. X-ray film normally gives lower resolution than does Kodak Tmax film, but its resolution is sufficient for identifying organs or regions of organs containing individual mRNAs, as is shown in Section III. Tmax film is generally preferred for precise localization of specific labeled cells, as is described in Section II. Because Tmax film is single-sided, one must be careful to expose the membrane to the emulsion side. One corner of the film is notched to allow the emulsion side to be identified in darkness. When the film is turned so that the notch is at its lower-right corner, the emulsion side is underneath. Develop Tmax film in total darkness in Tmax developer.

Exposure time depends on the signal intensity. In our experience, exposure time may be as short as 4 hr or as long as 4 days. The mRNA localization on the film can be observed with a hand lens or dissecting microscope. A disadvantage of tissue prints is that anatomy cannot be seen on the film. Therefore, it is important to compare the image on the film with the anatomy of the corresponding stained section in order to identify with certainty the tissue-level localization of the signal. Kodak Technical Pan 2415 is recommended for recording results.

Several kinds of controls can be used to check the specific hybridization on tissue prints: (a) Sense RNA probes can normally be used as a control of background for most mRNA localization, but (b) heterologous probes, such as those synthesized from vectors, and (c) probes or membranes treated with RNase before hybridization can also be used; also, (d) probes synthesized from housekeeping genes or from other genes in which the spatial distribution of mRNA is known can be used for positive controls. It is important to include controls in each experiment.

In situ hybridization shows both signal and anatomical information. Therefore, it is easy to determine the specific mRNA localization at the cellular level. Tissue print hybridization provides signal information on the film, but it does not provide precise anatomical information. However, by comparing the spatial distribution of the signal with the anatomy of the corresponding physical print and the corresponding stained section, one can conveniently localize the signal at the tissue level. When the localization of mRNAs is not certain, we recommend using in situ hybridization for comparison. If the antibody against the gene product of interest is available, immunohistochemical data on localization of the protein can also be used for comparison.

Resolution also depends on the size of the plant material used. In our experience, tissues having a diameter greater than 1.5 mm provide sufficient resolution to distinguish the signal in xylem from that in phloem. However, for tissues less than 1 mm in diameter, the spatial distribution of the signal generally cannot be accurately determined.

As tissue print hybridization continues to be used, new techniques will surely improve its resolution. One improvement would be to use nonradioactive labeled probes to directly detect the mRNA of interest on the membrane; e.g., one could use digoxigenin-labeled probes, visualized by alkaline phosphatase conjugated antibody against digoxigenin (Tautz & Pfeifle, 1989). This technique would allow the signal and physical print to be visualized on the membrane at the same time and would also produce a much sharper signal.

II. Tissue Distribution of Auxin-Induced mRNAs[1]

Tissue print hybridization, first described by Varner, Song, Lin, and Yuen (1989), uses a modified immunological tissue printing procedure that was initially developed by Cassab and Varner (1987). The tissue print hybridization procedure was modified by McClure and Guilfoyle (1989a, 1989b) for detecting moderately abundant mRNA transcripts with ^{35}S-labeled antisense RNA probes.

We have used tissue print hybridization to detect the organ and tissue distribution of small auxin-responsive mRNAs (SAURs) in whole seedlings and organ sections, both untreated and treated with auxin (McClure & Guilfoyle, 1989a, 1989b), and in hypocotyl sections undergoing gravitropic curvature (McClure & Guilfoyle, 1989a). In addition, we have used tissue printing to demonstrate that protein synthesis inhibitors, which cause an increase in SAUR abundance by a posttranscriptional mechanism, stimulate the accumulation of SAURs in the same tissues and organ regions as does auxin (Franco, Gee, & Guilfoyle, 1990). Ye and Varner (1991) have used tissue print hybridization to localize mRNAs that encode HRGPs and GRPs in developing soybean tissues.

The primary advantages of tissue print hybridization over *in situ* hybridization with fixed tissue sections are that tissue print hybridization is far less time consuming, is less expensive, and requires minimal technical expertise and equipment. Another advantage of tissue print hybridization is that it allows one to look at numerous tissue treatments (e.g., types of auxins and nonauxin analogs, dose–response, ki-

[1]This section was contributed by Tom J. Guilfoyle, Bruce A. McClure, Melissa A. Gee, and Gretchen Hagen.

netics of response) with a minimal amount of effort. All of the prints can be made on a single piece of nylon membrane, and after the prints have been made for each treatment, all subsequent procedures (i.e., staining, prehybridization, hybridization, and autoradiography) can be uniformly carried out.

Figure 7.1 shows tissue print and *in situ* hybridization of excised elongating soybean hypocotyl sections that were incubated in the presence or absence of 50 µM 2,4-D. Figure 7.2 is a tissue print hybridization of a 3-day-old etiolated soybean seedling treated with 2.5 mM 2,4-D for 1 hr, and Fig. 7.3 shows hypocotyls undergoing gravitropic curvature.

A. Materials

1. Nylon membranes (Zetaprobe, Bio-Rad)
2. 3-mm Whatman paper (Fisher)
3. Ultraviolet (UV) light source (300–320 nm)
4. Kapak or Scotchpak pouches
5. pIBI25 (International Biotechnologies) or some other suitable vector for synthesizing antisense or sense RNAs
6. Phenol:chloroform:isoamyl alcohol (25:24:1)
7. T7 RNA polymerase (Bethesda Research Laboratories) or other commercially available kit for synthesizing antisense or sense RNA probes
8. Placental ribonuclease inhibitor and RNase-free DNase (Promega)
9. ATP, CTP, GTP, dithiothreitol (DTT), sodium dodecyl sulfate, salmon sperm DNA, dodecyl A, yeast tRNA, polyvinyl-pyrrolidone, bovine serum albumin (BSA), Ficoll, and formamide (Sigma) [deionize the formamide before use by stirring for 30 min in AG501-X8 ion exchange resin (Bio-Rad) (see Sambrook, Fritsch, & Maniatis, 1989)]
10. ^{35}S-thio-UTP, >1200 Ci/mmol (New England Nuclear)
11. Sephadex G-50 (Pharmacia)
12. SSC: 0.15 M NaCl and 0.015 M sodium citrate
13. SSPE: 0.18 M NaCl, 0.01 M sodium phosphate (pH 7.4), and 1 mM EDTA
14. 50 X Denhardt's solution: 5 g Ficoll, 5 g polyvinylpyrrolidone, and 5 g BSA diluted to 500 ml with water

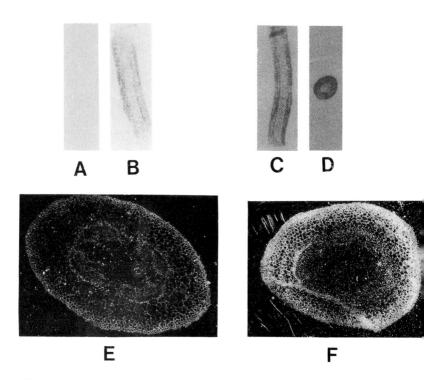

Figure 7.1 Excised elongating soybean hypocotyl sections were incubated in the presence or absence of 50 μM 2,4-D. After incubation, sections were cut longitudinally or cross-sectionally, and the cut surfaces were blotted onto a nylon membrane for tissue print hybridization or were fixed for *in situ* hybridization (Cox & Goldberg, 1988). A SAUR antisense RNA labeled with ^{35}S-thio-UTP was used as a hybridization probe. *In situ* hybridization with fixed tissue cross sections was carried out essentially as described by Cox and Goldberg (1988) with antisense ^{35}S-labeled SAUR probes (M. Gee, G. Hagen, & T. Guilfoyle, submitted). A–D are tissue print hybridizations. E and F are *in situ* hybridizations that show up under dark-field microscopy. A: hypocotyl longitudinal section incubated without 2,4-D for 1 hr; B: hypocotyl longitudinal section incubated with 2,4-D for 1 hr; C: hypocotyl longitudinal section incubated with 2,4-D for 2 hr; D: hypocotyl cross section incubated with 2,4-D for 2 hr; E: hypocotyl cross section incubated without 2,4-D for 1 hr; F: hypocotyl cross section incubated with 2,4-D for 1 hr. Little hybridization is detected in the section that was not treated with 2,4-D; however, in the sections treated with 2,4-D, hybridization is detected primarily in the epidermis and cortex. This tissue pattern of hybridization can be seen in both the longitudinal and cross sections. Tissue print hybridization provides a relatively accurate representation of the SAUR abundance in the soybean hypocotyl tissues because *in situ* hybridization with fixed tissue sections (E and F) reveals a pattern of hybridization that is almost identical to that seen with tissue prints.

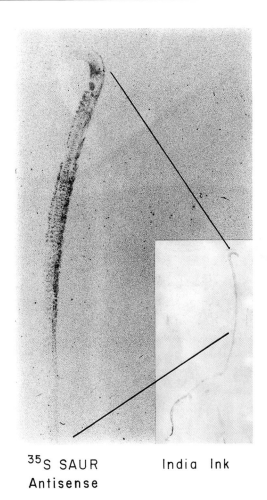

^{35}S SAUR   India Ink
Antisense

Figure 7.2 For tissue print hybridization a 3-day-old etiolated soybean seedling was treated with 2.5 mM 2,4-D for 1 hr, sprayed to runoff with 2,4-D, and placed in the dark for 1 hr before longitudinal sectioning. The inset is a reverse image of the tissue print stained with India ink. A SAUR antisense RNA labeled with ^{35}S-thio-UTP was used as a hybridization probe. Hybridization is detected primarily in the epidermis and cortex of the elongation zone of the hypocotyl, although a slight amount is also detected in the pith. Little, if any, hybridization is detected in the vascular tissue. Hybridization is much weaker in the meristematic or hook region and in the basal or mature region of the hypocotyl than in the elongation region. The organ region distribution of the SAURs detected with tissue print hybridization is in agreement with the distribution detected by Northern blot analysis using RNA purified from different organs and organ regions of the soybean seedling (McClure & Guilfoyle, 1987).

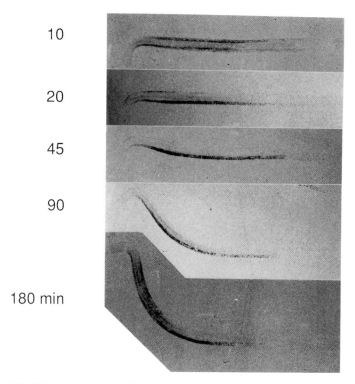

10

20

45

90

180 min

Figure 7.3 Tissue print hybridization was performed with soybean hypocotyl undergoing gravitropic curvature. Elapsed time after the hypocotyls were reoriented from the vertical to the horizontal is indicated. A SAUR antisense RNA labeled with ^{35}S-thio-UTP was used as a hybridization probe. During gravitropism, it is likely that endogenous auxin becomes more concentrated on the lower side of the hypocotyl than in the upper side. Tissue print hybridization reveals that the SAURs rapidly disappear from the top of the hypocotyl and become more abundant on the bottom of the hypocotyl during gravitropism. The tissue distribution of the SAURs is again primarily in the epidermis and cortex of the elongating portion of the hypocotyl. In this case, tissue printing is a relatively high-resolution method for studying the kinetics of SAUR abundance and distribution during gravitropic curvature.

15. India ink
16. Kodak XRP-5 x-ray film; Tmax 400, Technical Pan 2415, and Technical Pan 4415 photographic films; and Kodak Tmax developer, stop bath, and fixer

B. Printing Membrane

1. Place a nylon membrane over a single layer of dry 3-mm Whatman paper; the membrane and paper are not wetted before tissue printing. Wear vinyl medical gloves when handling the nylon membrane and blotting the tissue sections onto the membrane.

2. Prepare organ sections for printing by sectioning through the organ with a razor blade. Immediately press the freshly cut surfaces onto the nylon membrane, applying firm pressure above the sections with the index finger for 30–120 sec.

3. After printing, remove the organ sections from the nylon membrane with forceps, and air dry the membrane.

4. Evaluate the quality of the tissue prints by examining the printed nylon membrane under the UV light source. Poor quality will result if the organ sections are crushed, distorted, or unevenly blotted because of too much or too little pressure or if the tissue on the membrane is moved during printing. Each type of organ and tissue has its own unique characteristics, and the amount of pressure required to obtain even, consistent prints must be determined empirically. Large, firm organs (e.g., hypocotyls, stems, and cotyledons) are much easier to tissue print than are small, softer organs (e.g., floral parts, leaves, and roots). **Caution:** Ultraviolet radiation is dangerous, particularly to the eyes. To minimize exposure, make sure that the UV light source is shielded and wear protective goggles or a safety mask to block the UV light.

Dried prints can be kept at room temperature for several weeks without noticeable deterioration, but it is best to use the prints immediately or store them at 4°C in sealed Kapak or Scotchpak pouches.

C. Preparation of Antisense RNA Probe

1. Linearize full-length or partial cDNAs cloned into pIBI25 or some other suitable vector with appropriate restriction enzymes to allow production of antisense RNA *in vitro*.

2. Verify linearization of the cDNA clone by agarose gel electrophoresis.

3. Extract the linearized DNA with one volume of phenol:chloroform:isoamyl alcohol (25:24:1), and precipitate the DNA from the aqueous phase with two volumes of 95% ethanol at −80°C for 1 hr (Sambrook, Fritsch, & Maniatis, 1989).

4. Recover the precipitated DNA by centrifuging in a microfuge, and after decanting the supernatant, dry the DNA *in vacuo*. Suspend the dried DNA pellet in sterile, deionized water, and determine the concentration of the DNA by 260-nm absorption (Sambrook, Fritsch, & Maniatis, 1989).

5. Synthesize the antisense RNA in a 10-μl reaction mixture containing T7 RNA polymerase buffer, 0.3–0.7 μg of linearized DNA template, 10 U of T7 RNA polymerase, 10 mM DTT, 30–40 U of placental ribonuclease inhibitor, 1 mM ATP, CTP, and GTP, and 0.1–0.5 mCi of ^{35}S-thio-UTP. Dry the labeled UTP *in vacuo* immediately before using. Incubate the reaction mixture at 37°C for 40–60 min, and then add a second aliquot of T7 RNA polymerase (1 μl, 20 U) and continue incubation for an additional 40–60 min.

6. After incubation, add RNase-free DNase I (1 U) for 15 min to digest the DNA template.

7. Remove unincorporated nucleotides from the RNA by passing the reaction mixture through a 0.5-ml spin column (Sambrook, Fritsch, & Maniatis, 1989) of Sephadex G-50.

The antisense RNA probe may also be prepared with a commercially available kit that uses T7, T3, or SP6 RNA polymerase (e.g., Riboprobe Transcription System from Promega). A sense RNA probe should also be synthesized and used as a negative control in hybridization.

^{32}P-α-UTP or any other ^{32}P-labeled ribonucleoside triphosphate may be substituted for the ^{35}S-thio-UTP. ^{32}P-labeled probes are less expensive to synthesize and have higher energy (require less exposure time) than ^{35}S-thio-UTP. On the other hand, the ^{35}S-labeled probes provide greater resolution. The amount and specific activity of the ^{32}P- or ^{35}S-labeled ribonucleotide used to synthesize the antisense RNA probe that will detect its corresponding mRNA on a tissue print depend on the abundance of that mRNA in the printed tissue.

D. Hybridization of RNA to Tissue Prints

1. Prewash the dried tissue prints for 4–12 hr in 0.1–0.2 X SSC and 1% SDS at 65°C.
2. Perform prehybridization and hybridization at 68°C in 1.5 X SSPE, 1% SDS, 1% powdered milk, 0.5 mg/ml denatured sonicated salmon DNA, and 100 mM DTT. Generally, prehybridization is carried out for 12–16 hr, followed by hybridization with the antisense RNA probe at 5×10^7 cpm/ml in fresh buffer for 12–24 hr.
3. After hybridization, rinse the tissue prints briefly in 2 X SSC, 1% SDS, and 10 mM DTT, and then wash two more times for 30 min each at 42°C with gentle shaking. Wash two more times for 30 min each in 0.2 X SSC, 1% SDS, and 1 mM DTT at 65°C.

As an alternative to the prehybridization and hybridization buffer described, we have also successfully substituted a buffer consisting of 50% formamide, 5 X Denhardt's solution (Sambrook, Fritsch, & Maniatis, 1989), 6 X SSC, 2 mM EDTA, 0.1% SDS, 200 µg/ml poly A, 100 µg/ml yeast tRNA, and 70 mM DTT. Prehybridization and hybridization are carried out at 42°C.

E. Staining

1. Stain the tissue prints with India ink (Cassab and Varner, 1987) before autoradiography.

2. Rinse the prints in ice water until little or no streaking is observed when the membrane is lifted from the water, and then immerse in ice-cold India ink for 1–10 min.
3. Destain the prints by briefly rinsing in ice water and then rinsing several times in 0.2 X SSC and 1% SDS.

Although ink-stained images are not always uniformly stained, they are useful for comparing with the autoradiographic images. The ink-stained images provide anatomical detail that is not revealed in the autoradiograms, and this is important for interpreting the autoradiograms in terms of tissue-specific gene expression. The ink-stained images are also useful for verifying that the absence of a hybridization signal was not caused by ineffective blotting of the tissue sample. In our experience, many ink-stained images are not completely uniform but nevertheless yield relatively uniform hybridization signals over a particular tissue or organ region.

F. Autoradiography

We have evaluated several autoradiographic methods to find conditions that best display the tissue prints. For initial analysis of prints, we routinely use Kodak XRP-5 film exposed for 48 hr at $-70°C$. We have found that the grain quality of this film is superior to that of Kodak XAR-5 and SB-5. Because the ^{35}S radiation penetrates the plastic base of the films poorly, the images on the autoradiograms can be brightened by removing the photographic emulsion on the unexposed side. The emulsion can be removed by washing with household bleach and rinsing with water.

Although the preceding autoradiographic procedure can provide an acceptable hybridization signal, higher-resolution autoradiograms can be obtained by using photographic, rather than autoradiographic, films. We routinely use Kodak Tmax 400 film, which requires exposure times about five times as long as those of XRP-5 but produces a higher-quality image. With SAUR probes, most of the Tmax 400 autoradiograms require $\approx$10-day exposures at $-70°C$. The film is developed for 7–11 min at 24°C in Kodak Tmax developer, stopped for 30 sec in Kodak stop bath, and fixed for 5 min in Kodak fixer.

Note

We have tested other films, including Kodak Tri-X Pan, Tmax 3200, and Technical Pan 2415 or 4415. We have found that Tri-X Pan does not perform as well as Tmax 400 at the high film speeds used and that the Tmax 3200 is not substantially faster than Tmax 400 in our application. The highest-quality images have been obtained with Kodak Technical Pan 2415 (35-mm format) or 4415 (4 × 5-in. format). This film is substantially slower than Tmax 400; however, we have not directly compared the exposure times required for Tmax to those for Technical Pan. The extremely fine grain of Technical Pan films makes them superior if hybridization signals are strong; however, their slow speeds limit their application.

III. Localization of β-Conglycinin and Extensin mRNAs[2]

Tissue printing on nitrocellulose membrane was first developed to localize extensin in soybean seed with specific antibodies (Cassab & Varner, 1987). Tissue printing was further developed to localize extensin mRNA in soybean pods and α-subunit mRNA of β-conglycinin at different developmental stages of soybean seeds (Fig. 7.4) by using ^{32}P-labeled DNA probes (Varner, Song, Lin, & Yuen, 1989). The protocols presented here describe the tissue printing technique for localizing specific mRNAs in plant tissue with radiolabeled DNA and antisense RNA probes (Fig. 7.5). The results demonstrate that tissue print hybridization on membranes permits simultaneous examination of the pattern of gene expression and the morphology of the tissue, that it is especially suitable for large organs, and that it is fast and simple.

[2]This section was contributed by Yan-Ru Song and Joseph E. Varner.

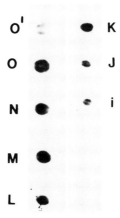

Figure 7.4 α-Subunit mRNA of β-conglycinin at different developmental stages of soybean seeds is localized by tissue print hybridization on nitrocellulose membrane. Hybridization was performed with the α-subunit gene of β-conglycinin (12-kb EcoR I fragment of soybean genomic DNA, Gmg 17.1, obtained from Dr. Roger Beachy, Washington University). The probe was labeled with [α-^{32}P]dCTP (1×10^7 to 2×10^7 cpm/ml). Soybean seeds (*Glycine max* cv. Provar) at different developmental stages were used. 0', mature yellow seed; 0, mature green seed more than 28 days after anthesis; N, 24–28 days; M, 21–23 days; L, 20–22 days; K, 19–21 days; J, 18–20 days; I, 17–19 days. Hybridization was performed at 42°C for 20 hr. The membrane was washed in 2 X SSC and 0.1% SDS at 42°C for 1 hr and in 0.5 X SSC and 0.1% SDS at 65°C for 1 hr. The filter was exposed to x-ray film for 6 hr.

TISSUE PRINT HYBRIDIZATION WITH DNA PROBE

A. Materials

1. Nitrocellulose membrane (Schleicher & Schuell)
2. Denatured DNA
3. [α-^{32}P]dCTP, 10 mCi/ml (New England Nuclear)
4. Klenow fragment of *E. coli* DNA polymerase I (Stratagene)
5. Sephadex G-50 (Pharmacia)
6. 4 X stock buffer: 200 mM Tris-HCl (pH 8.0), 20 mM MgCl$_2$, 20 mM DTT, 80 mM dATP, dGTP, and dTTP, 800 mM Hepes buffer, and 21.6 U/ml hexamer

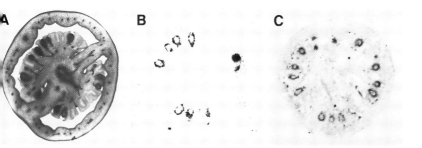

Figure 7.5 Extensin mRNA in tomato fruit 12 days after anthesis is localized by tissue print hybridization on nylon membrane. (A) A freehand section of tomato fruit was stained with toluidine blue. (B) A tomato fruit was cut in half, and the freshly cut surface was printed on the nylon membrane. The membrane was probed with ^{35}S-labeled extensin antisense RNA. The hybridization signal appears in the seed coat area of the tomato fruit. (C) A tomato fruit was cut in half, and the freshly cut surface was printed on the nitrocellulose membrane. The membrane was probed with antibodies against soybean seed coat extensin (Cassab & Varner, 1987). The soluble extensin is mainly found in the seed coat area.

7. Nick translation stop solution: 20 mg/ml blue dextran 2000, 0.1% SDS, and 50 mM EDTA

8. Prewashing solution: 50 mM Tris-HCl (pH 8.0), 1 M NaCl, 1 mM EDTA, and 0.1% SDS

9. Prehybridization solution: 40% formamide, 5 X SSC, 5 X Denhardt's solution, 0.1% SDS, and 0.25 mg/ml salmon sperm DNA

10. Kodak XAR-5 x-ray film

B. Preparation of DNA Probe by Random Oligonucleotide Primed Synthesis

1. Prepare the reaction mixture: 0.1–1.0 µg denatured DNA; 7.5 µl 4 X stock buffer, 4.0 µg BSA, 2–3 U Klenow fragment of E. coli DNA polymerase I, 1×10^7 to 2×10^7 cpm [α-^{32}P]dCTP. Dilute the reaction mixture to a total volume of 30 µl with water.

2. Incubate the reaction at room temperature for 2–4 hr.
3. Add nick translation stop solution to the reaction mixture. Load the probe onto a Sephadex G-50 spin column, and spin at 200 g until the dye marker separates completely. The unincorporated $[\alpha\text{-}^{32}P]dCTP$ remains in the syringe, and the labeled DNA is collected from the vial.

C. Procedure

1. Press a freshly cut section of tissue onto a nitrocellulose membrane for 20–30 sec with a gloved finger; it is important that the printing pressure be uniform from one sample to the next. After printing, carefully remove the section from the membrane with sharp forceps.
2. Bake the membrane in a vacuum oven at 80°C for 2 hr. The dried tissue prints can be stored for up to 2 mo at 4°C in a sealed plastic bag.

D. Hybridization and Autoradiography

1. Prewash the membrane in the prewashing solution on a shaker at 42°C for 2 hr.
2. Prehybridize the membrane in a meal bag on a shaker at 42°C for 2–4 hr.
3. Boil the radiolabeled probe for 5 minutes. Use enough probe to allow 10^7 cpm per sheet of nitrocellulose membrane (12 × 14 cm).
4. Hybridize the membrane in its storage bag with prehybridization solution and probe. Place on a shaker at 42°C for 16–20 hr.
5. Wash the membrane in 2 × SSC and 0.1% SDS at 42°C two times for 1 hr each.
6. Wash the membrane in 0.5 × SSC and 0.1% SDS at 65°C for 1 hr.
7. Using an intensifying screen, expose the membrane to x-ray film at −70°C for several hours to several days, and develop.

TISSUE PRINT HYBRIDIZATION WITH
RNA PROBE

A. Materials

1. Nylon membrane (Zetaprobe, Biorad)
2. $[\alpha\text{-}^{35}S]$rUTP, 12.5 mCi/ml (New England Nuclear)
3. RNase block II (Stratagene)
4. T_7 or T_3 RNA polymerase (Stratagene)
5. RNase-free DNase I (Stratagene)
6. Hybridization solution: 1.5 X SSPE, 1 X SDS, 1% BSA, 0.5 mg/ml denatured sonicated salmon DNA, and 100 mM DTT
7. 5 X transcription buffer: 200 mM Tris-HCl (pH 7.5), 30 mM $MgCl_2$, 10 mM spermidine, and 50 mM NaCl
8. Reaction mixture: 5 µl $[\alpha\text{-}^{35}S]$rUTP, 2 µl 100 mM DTT, 4 µl 5 X transcription buffer, 2 µl 10 mM NTP mixture (containing 10 mM ATP, 10 mM CTP, and 10 mM GTP), 1 µl linearized template DNA (0.1–0.5 µg); 1 µl RNase inhibitor; 1 µl T7 or T3 RNA polymerase (20 U/µl when pBluescript vector is used), and 4 µl water treated with diethyl pyrocarbonate
9. Sephadex G-50-100 (Pharmacia)
10. Kodak Tmax 400 film (4 × 5 in.), Tmax developer, stop bath, and fixer

B. RNA Probe Synthesis

1. Dry 0.5 mCi $[\alpha\text{-}^{35}S]$rUTP in a vacuum, and resuspend it in 5 µl water.
2. Incubate the reaction mixture at 37°C for 1 hr. Add 1 µl of RNA polymerase, and incubate for another hour. Digest the template DNA with 1 U DNase, and incubate at 37°C for 15 min.
3. Dilute the reaction mixture to 100 µl. Remove unincorporated nucleotides by passing the reaction mixture over a 1-ml Sephadex G-50-100 spin column.

C. Procedure

Make tissue prints as described for DNA probe except use Ze
taprobe nylon membrane instead of nitrocellulose membrane.

D. Hybridization and Autoradiography

1. Prewash the membrane for 2–4 hr in 0.2 X SSC and 1% SD$
   at 42°C.
2. Prehybridize in a sealed bag containing hybridization solutio
   for 12–16 hr at 65°C.
3. Hybridize with the probe at 2×10^7 to 5×10^7 cpm/ml fo
   16–36 hr in hybridization solution.
4. After hybridization, rinse the membrane in 2 X SSC, 1% SDS
   and 10 mM DTT, and then wash two or three times for 3(
   min each at 42°C with gentle shaking. Finally, wash the mem
   brane two times in 0.2 X SSC, 1% SDS, and 1 mM DTT fo
   30 min each at 65°C.
5. Expose the membrane to Tmax 400 film; the duration of ex
   posure depends on the intensity of the signal. Develop th
   film for 11 min at room temperature in Tmax developer, sto
   with Kodak stop bath for 30 sec, and fix for 5 min in Koda
   fixer. Wash the film in running water. Observe the result
   under a dissecting microscope.

IV. Localization of HRGP and GRP mRNAs in Developing Soybean Stem

HRGPs and GRPs are two classes of cell wall structural proteins i
higher plants (Cassab & Varner, 1988). Much evidence indicates tha
these cell wall proteins have tissue-specific distributions and that thei
gene expression is developmentally regulated. We have used tissu

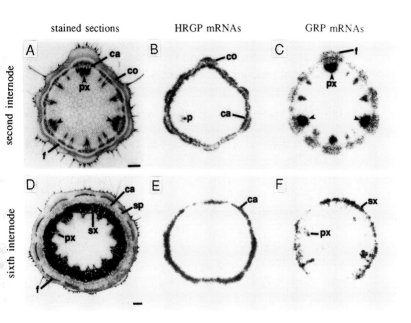

Figure 7.6 HRGP mRNAs and GRP mRNAs in developing soybean stem are localized by tissue print hybridization on nylon membrane. Carrot HRGP genomic DNA, pDC5A1 (Chen & Varner, 1985b), and bean GRP 1.8 genomic DNA, a kind gift from Drs. Beat Keller and Christopher J. Lamb (Keller, Sauer, & Lamb, 1988), cloned into pBluescript vector (Stratagene) were used to synthesize ^{35}S-labeled antisense RNA probes by using T$_7$ or T$_3$ RNA polymerase. Scale bar represents 300 μm. Abbreviations: ca, cambium; co, cortex; f, primary phloem; p, parenchyma; px, primary xylem; sp, secondary phloem; sx, secondary xylem. From "Tissue-Specific Expression of Cell Wall Proteins in Developing Soybean Tissues" by Z.-H. Ye and J. E. Varner, 1991, *Plant Cell* **3**, pp. 23–37.

print hybridization on membranes as described by McClure and Guilfoyle (1989a, 1989b) and Varner, Song, Lin, and Yuen (1989) to show that the gene expression of HRGPs and GRPs (Fig. 7.6) is temporally and spatially regulated in developing soybean stems (Ye & Varner, 1991).

A. Materials

1. Nylon membrane (Zetaprobe, Bio-Rad)
2. 10 X transcription buffer: 400 mM Tris-HCl (pH 7.5), 60 mM MgCl$_2$, 20 mM spermidine-HCl, and 50 mM NaCl
3. Transcription reagents (at room temperature): 0.5 μg linearized DNA template, 1 μl 100 mM DTT, 2.5 μl 10 X transcription buffer, 1 μl 10 mM rATP, 1 μl 10 mM rCTP, 1 μl 10 mM rGTP, 1 μl 1 mM rUTP, 10 U RNase block II (Stratagene), 50 mCi [α-^{35}S]rUTP (New England Nuclear), 10 U T7 or T3 bacteriophage DNA-dependent RNA polymerase (Stratagene), and enough water to bring the final volume to 25 μl
4. RNase-free DNase I (Stratagene)
5. Sephadex G-50 (Pharmacia)
6. Hybridization solution: 2 X SSC, 1% SDS, 5 X Denhardt's solution, 0.1 mg/ml salmon sperm DNA, and 10 mM DTT
7. Washing solution I: 0.2 X SSC and 1% SDS
8. Washing solution II: 2 X SSC and 0.1% SDS
9. Washing solution III: 0.2 X SSC and 0.1% SDS
10. Kodak Tmax 400 film (4 × 5 in.), Tmax developer, stop bath fixer, and Technical Pan 2415 film
11. One-month-old soybean plant

B. Tissue Printing

1. Put six layers of Whatman No. 1 paper on a plastic plate. Put one sheet of photocopy paper on the Whatman paper, and place the nylon membrane on it.
2. Use a double-edged razor blade to freehand cut a 1 mm-thick section from the soybean stem, and use forceps to carefully transfer the freshly cut section onto the membrane. Do not move the section after it is transferred onto the membrane.
3. Put four layers of Kimwipes on the section, and press the section gently and evenly for 15–20 sec with one finger.
4. Remove the Kimwipes and the section carefully with forceps and either keep the entire section or cut a thin section from

it for anatomy comparison. The additional section can be stained with toluidine blue for observation of anatomy.

5. Bake at 80°C for 2 hr, after which the printed membrane is ready for hybridization.

C. Preparation of ^{35}S-Labeled RNA Probe

1. Incubate the transcription reagent solution at 37°C for 1–2 hr.
2. Add 1 μl of RNase-free DNase I (1 mg/ml). Incubate the reaction for 15 min at 37°C.
3. Add 50 μl of water, and purify the RNA by Sephadex G-50 column chromatography or ethanol precipitation (Sambrook, Fritsch, & Maniatis, 1989).

D. Hybridization, Autoradiography, and Data Analysis

1. Wash the membrane in washing solution I at 65°C for 4 hr.
2. Prehybridize in hybridization solution at 68°C for 2 hr.
3. Hybridize with radiolabeled probe, 1×10^6 to 5×10^6 cpm/ml, at 68°C for 20 hr.
4. Wash in washing solution II three times at 42°C for 20 min each.
5. Wash in washing solution III two times at 65°C for 30 min each.
6. Briefly wash in 0.2 X SSC at room temperature, and air dry.
7. Expose the membrane to Tmax 400 film at room temperature and develop the film as follows: 10–15 min in Tmax developer, 30 sec in stop bath, 5 min in fixer, and 10 min in water. Air dry.
8. Observe the autoradiographic results on the film under a dissecting microscope. Photographically record the results on Kodak Technical Pan 2415 film.

V. Tissue-Specific Localization of Moderately Abundant mRNAs[3]

Tissue print hybridization for mRNA detection is rapid and simple and gives reasonably good resolution at the tissue or organ level. McClure and Guilfoyle (1989b) reported using tissue printing to localize auxin induced mRNAs in soybean seedlings. We have adapted their proce- dure for localizing a moderately abundant cellulase mRNA in the leaf abscission zones of beans (Fig. 7.7). Many modifications of the proce dure were attempted, and the protocol describes the procedure that gave the best and most reproducible results.

A. Materials

1. Nylon membrane (Hybond N, Amersham)
2. 20 X SSPE: 3.6 M NaCl, 0.2 M Na_2HPO_4 (pH 7.4), and 20 mM EDTA
3. Hybridization buffer: 2 X SSPE, 2 X Denhardt's solution (0.2 g Ficoll, 0.2 g polyvinylpyrrolidone, and 0.2 g BSA brought to a final volume of 500 ml with water), 100 µg/ml denatured sonicated salmon sperm DNA (from 10 mg/ml stock), and 0.1% SDS (from 10% SDS stock)
4. Kodak XAR-5 x-ray film

Note

Double-distilled water used to prepare solutions should be treated with diethyl pyrocarbonate at 0.05% and then autoclaved. This treatment destroys RNase activity in the water.

B. Procedure

1. Refrigerate the organ at 4°C for several minutes; this is be lieved to reduce RNase activity released on cutting the tis

[3]This section was contributed by Mark L. Tucker.

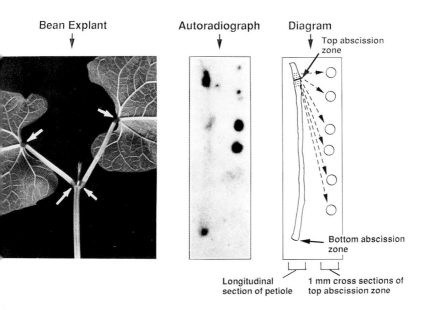

| Bean Explant | Autoradiograph | Diagram |

Figure 7.7 9.5 cellulase mRNA in abscission is detected by tissue print hybridization. The photograph labeled "Bean Explant" shows a 12-day-old bean explant before ethylene treatment (primary and trifoliate leaves were removed before treatment). White arrows point to both proximal and distal pulvini, where abscission fractures will form after exposure to ethylene. The top (distal) abscission zone forms at the base of the laminar pulvinus. The bottom abscission fracture line forms at the base of the petiolar pulvinus next to the stem. The "Diagram" shows the tissue sections that were printed onto the membrane used for the autoradiograph shown. The longitudinal section of the petiole includes only the distal side of the petiolar (proximal) abscission zone and both sides of the laminar (distal) abscission zone. The 1-mm cross sections to the right of the longitudinal section were from the top (distal) abscission zone of another explant from the same batch of ethylene-treated explants. The probe was a ^{32}P-labeled RNA transcript of the bean's abscission cellulase cDNA clone pBAC10 (Tucker & Milligan, 1991). From "Bean Leaf Abscission: Tissue-Specific Accumulation of a Cellulase mRNA" by M. L. Tucker, S. L. Baird, and R. Sexton (1991). *Planta* **186**, 52–57.

sue. Handle the tissue and hybridization membrane with gloves to reduce RNase contamination.

2. Place a dry, clean nylon membrane on blotting paper.
3. Dip a double-edged razor blade into cold 10 X SSPE, and slice the tissue.

4. Immediately place the cut surface down onto the membrane
   and press lightly to bring the cut surface into contact with
   the membrane. It is important not to press the tissue hard
   enough to rupture uncut cells.
5. Allow the membrane to air dry.
6. Expose the membrane to UV irradiation (we use 5 min on a
   UV transilluminator).
7. Prepare the probe (we use ^{32}P-labeled RNA transcripts to
   optimize the hybridization signal).
8. Incubate the printed membrane with gentle rocking in hybridization buffer for 3–16 hr at 50°C, replace the buffer with
   fresh hybridization buffer containing the labeled RNA probe
   and incubate another 16 hr at 50°C.
9. Wash the membrane two times for 15 min each in 1 X SSPE
   and 0.1% SDS at room temperature and two times for 30
   min each in 0.1 X SSPE and 0.1% SDS at 65°C.
10. For the ^{32}P-labeled probe, expose the still moist membrane
    to preflashed Kodak XAR-5 x-ray film at –80°C with an in-
    tensifying screen. If the membrane has not been allowed to
    dry, it may be washed at higher stringencies to reduce
    background or may be incubated with RNase A (see the
    following).

Note

Better resolution may be obtained with a ^{35}S-labeled probe and
different films (see McClure & Guilfoyle, 1989b, and Sections I and II,
this chapter). Membranes may be incubated with RNase A (50 μg/ml
RNase A in 0.5 M NaCl, 10 mM Tris, pH 7.0, and 1 mM EDTA for 30
min at 37°C) after hybridization to reduce the signal from nonspecific
binding of single-stranded RNA transcripts (Cox & Goldberg, 1988).
Double-stranded RNA resulting from specific hybridization of the labeled probe is resistant to degradation by RNase A. Although treating
prints with RNase A may be an important control for assuring that the
autoradiographic signal is from RNA-specific hybridization, the treatment reduces signal intensity somewhat, which necessitates longer film
exposure.

VI. Localization of Pectin Methylesterase mRNA in Tomato Fruit[4]

As a tomato fruit ripens, biochemical reactions involved in color, flavor, and texture changes take place. Polygalacturonase (PG), a major cell wall protein, is induced in ripening tomato fruit, and it has been suggested that PG is a factor in fruit softening. Immunotechniques (see Chapter 4) have shown that the appearance of PG in tomato fruit is linked to maturation and ripening (Tieman & Handa, 1989). This protocol localizes mRNA encoding another enzyme related to the cell wall PG, pectin methylesterase (PME) (Fig. 7.8). This enzyme splits the methyl group of methylated polygalacturonates, a step necessary for PG activity.

A. Materials

1. 20 X SSC: 3 M NaCl and 0.3 M sodium citrate (pH 7.0)
2. Nitrocellulose membrane saturated with 20 X SSC
3. 5 X Denhardt's solution: 0.1% BSA, 0.05% Ficoll, and 0.1% polyvinylpyrrolidone
4. 0.1% SDS
5. Hybridization solution: 6 X SSC containing 5 X Denhardt's solution, 0.1% SDS, and 50 µg/ml denatured herring sperm DNA
6. PME cDNA clone labeled by random priming with [α-^{32}P]dCTP (see Section III)
7. Kodak XAR-5 film and intensifying screens

B. Procedure

1. Cut each fruit in half with a sharp knife, and place the cut surface onto the nitrocellulose membrane previously saturated with 20 X SSC. Apply light pressure when the fruit is placed

---

[4]This section was contributed by R.W.F. Harriman, Denise M. Tieman, and Avtar K. Handa.

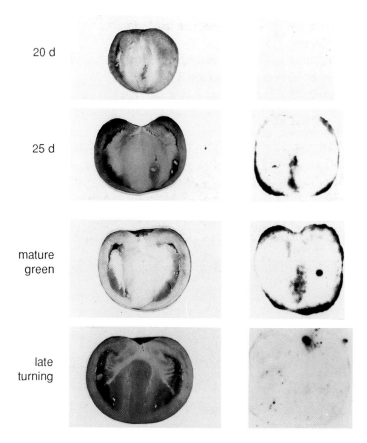

20 d

25 d

mature
green

late
turning

Figure 7.8 Pectin methylesterase (PME) mRNA during tomato fruit development is localized with tissue printing. The fruits on the left are 20 days old, 25 days old, mature green, and in the late turning stage; their respective tissue prints are on the right. PME mRNA begins to accumulate in the outer pericarp and columella regions by 25 days, reaches a maximum between 30 days and the mature green stage, and declines thereafter with fruit ripening.

on the membrane, and let the fruit remain on the membrane for 5 min.

2. Cross link the transferred RNA to the membrane by baking in an 80°C vacuum oven for 2 hr.

3. Prehybridize for 4 hr at 65°C in hybridization solution.

4. Add an insert of PME ^{32}P-mRNA clone, and hybridize for 48 hr at 65°C.

5. Wash the membrane three times for 1 hr each at 65°C in 1 X SSC containing 0.1% SDS.

6. After drying, expose the membrane to Kodak XAR-5 film at −80°C in the presence of two intensifying screens.

CHAPTER **8**

Detection and Localization of Plant Pathogens

Curtis A. Holt[1]

Department of Biology
Washington University
. Louis, Missouri

Overview

The determination of whether a plant is infected with a plant pathogen can be accomplished by a number of different methods. First, visual observation of the plant can reveal a pattern of pathogenic symptoms (such as chlorosis, necrosis, or deformity of growth) indicative of a particular pathogen. However, many pathogens induce visually identical

Additional contributions to this chapter have been made by Roger N. Beachy, Y. H. Hsu, N. S. Lin, and H. T. Hsu.

[1]*Current affiliation:* Division of Plant Biology, The Scripps Research Institute, La Jolla, California.

symptoms, precluding a precise identification of the causal agent. In addition, many pathogens are asymptomatic within a given host. Second, biological assay of infected tissues, such as transfer of tissue extracts onto indicator hosts, can indicate which tissues are affected and may indicate the identity of the pathogen involved. This method is limited by the similarity of symptoms induced on indicator plants by many different pathogens.

A third method used to determine the presence and identity of a plant pathogen involves the molecular analysis of the proteins or nucleic acids associated with the disease. The use of techniques such as protein gel electrophoresis, Western blotting, and nucleic acid hybridization have made precise identification and quantification of plant pathogens possible. However, such techniques are labor intensive and require large initial investments in equipment and reagents. In addition, neither the tissue localization of pathogens nor the kinetics of pathogenesis can be adequately addressed by such methods.

Tissue printing is a rapid and simple technique that has none of the disadvantages inherent in the methods just described. Tissue sections prepared from various plant parts (leaf, petiole, stem, root, bulb, etc.) are simply placed in direct contact with a nitrocellulose membrane for several seconds, and the resulting imprint is probed with immunoreagents specific for a particular pathogen. The antigenic reaction is then directly visualized on the membrane using protocols developed for immunoblots (Blake, Johnston, Russell-Jones, & Gotschlich, 1981). Navot, Ber, and Czosnek (1989) described one of the first applications of this procedure (a "squash blot") whereby tomato leaves, roots, stems, flowers, and fruits were squashed onto a nylon membrane with a glass rod or pen. The blots were subsequently probed with an antibody specific for tomato yellow leaf curl geminivirus to determine which plants and tissues were infected.

The tissue print method of detecting and localizing plant pathogens has many advantages. It is extremely simple and requires a minimum of equipment and reagents, and numerous samples can be examined for the presence of a pathogen in a relatively short time. Subliminal and early-stage infections without external symptoms can be readily detected. In addition, the specific nature of the antibody probes used can precisely identify the pathogen involved.

Probably the greatest strength of the tissue print method when applied to the analysis of plant pathogenicity is its potential to precisely localize the extent of the infection and the tissues involved. Assays of crude extract homogenates prepared from infected tissue cannot address these aspects of infection. For example, analysis of stem sections of plants infected with phloem-limited viruses indicate that viral antigen is restricted to the vascular tissues (see Fig. 8.2). Also, the timing and morphogenesis of the various stages of infection can be analyzed in detail by using tissue printing; pathogen antigens can often be detected extremely early in infection before any disease symptoms are manifested (Wisniewski, Powell, Nelson, & Beachy, 1990). Tissue prints prepared from simultaneously infected plants can be harvested and assayed at different times after inoculation to examine the pattern of disease development.

The following protocols represent some of the most recent applications of the tissue print technique to the detection and localization of plant pathogens.

I. Detection and Localization of Plant
Virus Antigens[2]

This basic protocol for preparing tissue prints to detect and localize the tissue distribution of phytopathogenic viruses details the imprinting procedure and the subsequent immunoblotting technique. Cross-sectional tissue specimens are prepared by hand and placed in brief contact with nitrocellulose membranes. A primary antibody prepared in rabbits or mice against a specific virus is reacted with the blot, and a secondary alkaline phosphatase (AP) anti-species antibody is then added. Antibody binding (and hence antigenic reaction) is visualized by using the substrate 5-bromo-4-chloro-3-indolyl phosphate (BCIP). After removal of the phosphate group by AP, a progressive oxidation

---

[2]This section was contributed by Curtis A. Holt and Roger N. Beachy.

reaction results in the formation of blue, insoluble 5,5'-dibromo-4,4' dichloroindigo, which precipitates on the nitrocellulose. This reaction is facilitated by adding the oxidation catalyst nitro blue tetrazolium (NBT).

Prints of cross sections from transgenic *Nicotiana tabacum* expressing the tobacco mosaic virus movement protein gene and infected with either wild-type tobacco mosaic virus or a movement defective clone of the virus are shown in Fig. 8.1, and the main stem of a *Physalis floridana* plant infected with potato leaf roll luteovirus is shown in Fig. 8.2.

A. Materials

1. Nitrocellulose membrane, 0.45–μm pore size (Schleicher & Schuell)
2. Bovine serum albumin (BSA), enzyme-linked immunosorbent assay (ELISA) grade (Sigma)
3. AP-conjugated goat anti-rabbit antibody (Promega) or goat anti-mouse antibody (Promega)
4. NBT substrate: 50 mg/ml NBT (Sigma) in 70% dimethylformamide
5. BCIP substrate: 25 mg/ml BCIP (Sigma) in dimethylformamide
6. Tween Tris buffer saline (TTBS): 10 mM Tris-HCl (pH 8.0), 150 mM NaCl, and 0.05% Tween-20
7. Blocking solution: 1% BSA in TTBS
8. AP buffer: 100 mM Tris-HCl (pH 9.5), 100 mM NaCl, and 5 mM MgCl$_2$
9. Color development solution: For every 10 ml solution, add 33 μl NBT substrate to 10 ml AP buffer, mix, add 66 μl BCIP substrate, and mix again; protect solution from strong light and use within 60 min

B. Procedure

1. Boil the nitrocellulose sheets in distilled water for 3 min.
2. Soak the sheets in 0.2 M CaCl$_2$ for 30 min at room temperature, and then blot dry on Whatman paper.

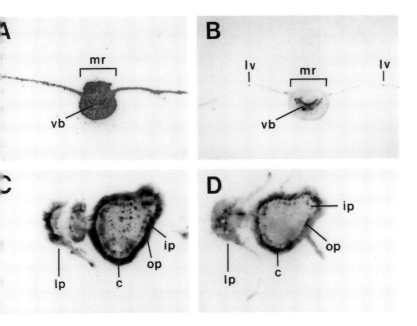

Figure 8.1 Prints were prepared from cross sections of transgenic tobacco plants expressing the tobacco mosaic virus (TMV) movement protein gene and infected with either wild-type TMV or a movement-defective TMV clone. The blots were prepared 10 days after inoculation and were probed with anti-TMV primary antibody: (A) section (3 cm long) from leaf (seventh leaf above inoculated leaf) of young wild-type plant infected with TMV; (B) section (3 cm long) from leaf (seventh leaf above inoculated leaf) of young plant infected with movement-defective TMV; (C) apical stem section from plant infected with wild-type TMV; (D) apical stem section from plant infected with movement-defective TMV. Abbreviations: c, cortex; ip, inner phloem; lp, leaf primordium; lv, lateral vein; mr, midrib; op, outer phloem; vb, vascular bundle. From "*In vivo* Complementation of Infectious Transcripts from Mutant Tobacco Mosaic Virus cDNAs in Transgenic Plants" by C. A. Holt and R. N. Beachy. (1991) *Virology.* **181**, 109–117.

3. Hand-cut a thin (≤1 mm) section of fresh tissue with a new razor blade.

4. Place the section on a pretreated nitrocellulose sheet, blot with hand pressure for 10–20 sec, and remove the tissue.

5. Wrap the blot in plastic film or aluminum foil and store at 4°C until developed.

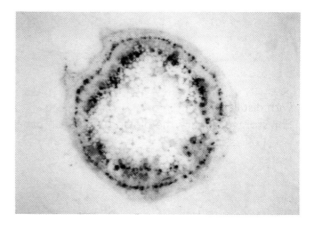

Figure 8.2 A tissue print was made from a cross section of the main stem of *Physalis floridana* plant infected with potato leaf roll luteovirus (PLRV). The blot was prepared 10 days after inoculation and was probed with anti-PLRV primary antibody. Viral antigen is clearly visible in the inner and outer phloem rings.

6. During the subsequent steps, do not allow the nitrocellulose to dry out. Perform all reactions at room temperature with gentle shaking in a shallow container that is slightly larger than the blot.

7. Float the blot on TTBS until it is evenly wet, submerge it, and rinse briefly in the same buffer.

8. To block excess protein binding sites, decant the buffer and incubate the blot in blocking solution for 30 min.

9. Bind the primary antibody: Replace the blocking solution with TTBS containing the appropriate dilution of primary antibody (typically, 1:5000 to 1:10,000), and incubate for 30 min.

10. Wash the blot in TTBS three times for 5–10 min each to remove unbound antibody.

11. Bind the anti-IgG AP conjugate: Transfer the blot to TTBS containing the appropriate anti-IgG AP conjugate (1:10,000 dilution of the Promega product is recommended), and incubate for 30 min.

12. Wash as in step 10.

13. Start the color reaction: Blot the nitrocellulose damp-dry on Whatman paper, and transfer to the color development solution. Protect from strong light. Reactive areas will turn purple, usually within 15 min.
14. Stop the color development by rinsing the blot several times with distilled water.
15. For storage, air dry the blot on Whatman paper. The color will fade slightly but can be restored by moistening with water. Protect the developed blot from strong light.

Notes

1. Large amounts of viral antigen present in the tissue section may cause "bleeding" of the colored product. To dilute the amount of antigen transferred to the blot, rinse the freshly cut tissue section in distilled water for 3 sec and blot on absorbent tissue before applying to the nitrocellulose. Alternatively, blot the fresh tissue section successively on several areas of the nitrocellulose to serially dilute the antigen transferred.
2. For relatively low levels of viral antigen in infected tissue, the use of glass-fiber filters results in more antigen being deposited on the blot. The Millipore GS filter (0.22 mm) is particularly good. It is processed exactly the same as nitrocellulose.

II. Detection of Plant Virus and Mycoplasmalike Organism Antigens[3]

A slight modification of the first protocol details variations in the immunoblotting procedure, including direct and indirect antibody reactions and the use of biotinylated antibody probes. This method has successfully been used to detect mycoplasmalike pathogens of plants

[3]This section was contributed by Y. H. Hsu, N. S. Lin, and H. T. Hsu.

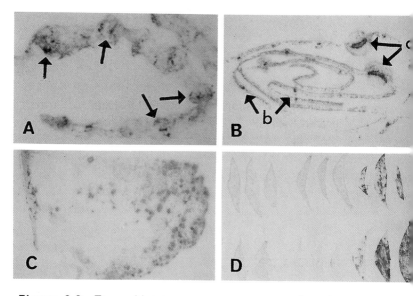

Figure 8.3 Tissue blot immunoassays were used to detect virus an mycoplasmalike organism infections in plants. (A) Localization of barley yellow dwa luteovirus antigens (arrows) in phloem of barley leaves with mouse monoclon: antibodies; (B) localization of tomato big bud inducing mycoplasmalike organisms i midrib phloem cells and secondary veins of leaves of infected periwinkle with mous monoclonal antibodies; (C) localization of bean yellow mosaic potyvirus antigens in dormant corm of gladiolus with mouse monoclonal antibodies; (D) detection c symptomless carlavirus infection in lily bulb scales with rabbit potyclonal antiser: From "Immunological Detection of Plant Viruses and a Mycoplasma-like Organism b Direct Tissue Blotting on Nitrocellulose Membranes" by N. S. Lin, Y. H. Hsu, an H. T. Hsu, (1990). *Phytopathology* **80,** 824–828.

(Fig. 8.3) (Lin, Hsu, & Hsu, 1990) and viral pathogens (Hsu & Lawson in press).

A. Materials

1. Nitrocellulose membranes
2. Phosphate-buffered saline (PBS): 20 mM K$_2$HPO$_4$ and 0.15 M NaCl (pH 7.4)
3. Blocking buffer: 2% BSA in PBS
4. Tween PBS (TPBS): PBS containing 0.05% Tween-20

5. Detection system:
    a. Virus-specific antibodies conjugated with AP
    b. Virus-specific primary antibodies (raised in rabbit or mouse) and secondary antibodies (goat anti-rabbit immunoglobulins for rabbit antisera or goat anti-mouse immunoglobulins for mouse monoclonal antibodies) conjugated with AP
    c. Biotinylated virus-specific antibodies and avidin–AP conjugate (Sigma)
6. Substrate solution: 14 mg NBT and 7 mg BCIP in 40 ml substrate buffer of 0.1 M Tris, 0.1 M NaCl, and 5 mM MgCl$_2$ (pH 9.5)

B. Procedure

1. Excise tissues (leaves, petioles, stems, flower buds, emerging shoots, corms, bulbs, etc.). For thin tissues, such as leaves, roll them into a tight core, hold in one hand, and cut with a new razor blade by using a steady motion with the other hand to obtain a single-plane cut surface.
2. Press the newly cut surface onto a nitrocellulose membrane with a firm but gentle force.
3. Block the tissue blot by immersing the membrane in blocking buffer for 30–60 min at room temperature.
4. Wash the membrane three times in TPBS, 10 min each time.
5. Immunological procedures:
    a. For direct immunological detection, incubate the tissue blot for 1–2 hr at room temperature with AP-labeled virus-specific antibodies diluted in PBS.
    b. For indirect immunological procedures, incubate the tissue blot for 1–2 hr at room temperature with virus-specific primary antibodies diluted in PBS. Wash as in step 4, then incubate with enzyme-labeled species-specific secondary antibodies for 1–2 hr at room temperature. Indirect immunological procedures can be modified by replacing the enzyme-labeled species-specific secondary antibodies with enzyme–protein A conjugates.

Figure 8.4 Trifoliate leaves from soybean plants were kept at −70°C for 10 min a returned to room temperature before press blotting. Pressure at 70 kg/cm² w applied for 1 min with a hydraulic press. The blots were incubated with a monoclor antibody to soybean mosaic virus coat protein. Antibody binding was detected using protein A–coupled colloidal gold particles followed by silver-sta enhancement. The light-colored leaf press blot (bottom) represents uninoculated l

(continu

   c. For immunological detection with biotinylated virus-spe-
cific antibodies, react the tissue blot for 1–2 hr at room
temperature with biotinylated antibodies diluted in PBS,
and wash as in step 4 before incubation for 1–2 hr at
room temperature with avidin–enzyme conjugate in PBS.

6. Wash as in step 4.
7. Immerse the tissue blot in substrate solution for color
development.
8. Quickly rinse the tissue blot with distilled water.
9. Stop the reaction by immersing the tissue blot in a solution
containing 0.01 M Tris and 1 mM EDTA (pH 7.5), two
changes for 10 min each.
10. Dry the membrane in a dust-free environment.

V. Localization of Plant Viruses in Leaf Tissue by Press Blotting[4]

Press blotting is a novel method that can be used to determine virus
location and distribution in entire leaves, allowing disease symptoms to
be analyzed in relation to sites of viral infection (Fig. 8.4). Press blotting
should also be useful to plant developmental biologists for locating the
distribution of specific plant proteins or mRNAs. Although the resolution of press blotting is not as great as that obtained by thin-section
tissue printing, the technique should be of general use to plant molecular biologists.

[4]This section was adapted from "Plant Virus Localization in Leaf Tissue by Press
Blotting" by L. M. Mansky, R. E. Andrews, Jr., D. P. Durand, and J. H. Hill (1990). *Plant
Molecular Biology Reporter* **8**, 13–17.

---

tissue; the dark-stained leaf press blot (top) represents virus-infected leaf tissue.
From "Plant Virus Localization in Leaf Tissue by Press Blotting" by L. M. Mansky,
R. E. Andrews, Jr., D. P. Durand, and J. H. Hill (1990). *Plant Molecular Biology Reporter*
8, 13–17.

A. Materials

1. Hydraulic press (model B, Fred S. Carver Inc.).
2. Nitrocellulose.
3. Tris buffer saline (TBS): 15 mM Tris and 130 mM NaCl, adjusted to pH 7.2 with HCl.
4. Blocking buffer: TBS containing 1.5% gelatin.
5. Monoclonal antibody to soybean mosaic virus (SMV) coat protein.
6. Protein A–colloidal gold solution (Bio-Rad).
7. Colloidal gold enhancement kit (Bio-Rad).
8. Soybean plants (*Glycine max* cv. Williams '82) (soybean leaf tissue is not very amenable to squash blotting). Maintain plants in a growth chamber at 22°C with a day length of 1 hr. Before development of trifoliate leaves inoculate primary leaves with an isolate of SMV. Sample trifoliate leaves 10 days after inoculation. As a control, sample soybean trifoliate leaves from uninoculated plants. Either immediately use the samples for the press blot or treat them at −70°C for 10 min and return them to room temperature before press blotting.

B. Press Blotting

1. Place two pieces of Whatman No. 1 filter paper below piece of dry nitrocellulose. The filter paper absorbs excess sap that is forced through the nitrocellulose.
2. Place leaf samples onto the dry nitrocellulose with the bottom surface of the leaf in direct contact with the membrane.
3. Wrap the sandwich of leaves and nitrocellulose in plastic food wrap.
4. Place the sandwich in the hydraulic press, and apply 70 kg/cm^2 for 1 min.
5. Disassemble the sandwich, and allow the nitrocellulose to air dry.

C. Detection of Virus

1. To eliminate nonspecific binding of detecting antibodies, incubate the press blot in blocking buffer for 30 min at 37°C.
2. Incubate the blot for 15 hr at 37°C in blocking buffer containing a 1:2000 dilution (of ascites fluid) containing monoclonal antibody to SMV coat protein.
3. Wash the blot in TBS for 30 min.
4. Place the blot in undiluted protein A–colloidal gold solution, and incubate until color develops.
5. Enhance the color by using the colloidal gold enhancement kit.

Note

Press blotting of virus-infected leaf tissue is a simple and efficient method for determining virus location in leaf tissue. A pressure of 70 kg/cm^2 for 1 min yields optimal extrusion of plant sap from leaf tissue. Lower background is observed with nitrocellulose matrices than with nylon-based membranes. We have found that the best leaf image is obtained when the leaves are frozen for 10 min at −70°C and then thawed before blotting. For applications in which the location of enzymatic activity is desired, leaf tissue can be blotted without prior treatment; however, less detail is seen when the soybean leaf tissue is immediately blotted. We have also tried electrophoretic transfer as an alternative to high pressure for blotting leaf material onto nitrocellulose paper, but the image resolution is lower than that obtained by press blotting.

CHAPTER 9

Tissue Prints of Animal Tissues

Philip D. Reid

Department of Biological Sciences
Smith College
Northampton, Massachusetts

I. Overview

Tissue printing on surfaces containing substrate molecules was first demonstrated with animal cells. The early work of Daoust (1957, 1965) is particularly worth noting, and some of his results are reported in Chapter 1. Moreover, the highest-resolution tissue prints yet

Additional contributions to this chapter have been made by Rafael F. Pont-Lezica.

TISSUE PRINTING
Copyright © 1992 by Academic Press, Inc.

139

achieved have been from animal cells. Gaddum and Blandau (1970) clearly showed the entire morphology of sperm cells that were printed onto gelatin surfaces (see Fig. 1.2). However, because most animal tissues are softer than most plant tissues, the use of blotting membranes for physical and chemical printing of animal tissue has not been fully pursued. It seems likely that the techniques that have been successful for plant tissues could be easily modified for studying processes in harder animal tissues, such as the mineralization of bone.

Although animal tissues do not generally lend themselves to physical printing, chemical information is transferred to membranes, so that the distribution of both proteins and nucleic acids can be determined. In what might be considered a variation of tissue printing, Barres has shown that individual nerve cells with their processes intact can be blotted onto nitrocellulose. The cells that have been transferred in this way can then be further analyzed (Barres, Koroshetz, Chung, & Corey, 1990). A similar technique involving purifying stereocilia from saccular hair cells of bullfrogs by cell adhesion to nitrocellulose has also been reported (Shepherd, Barres, & Corey, 1989). The proteins in the purified cells can then be studied. The technique, which has been referred to as "bundle blot" purification, will probably be useful for other tissues.

Brain blotting (Hernandez Bronchud, Webb, & Esiri, 1988), in which frozen sections of brain tissue are printed onto nylon membranes, is more analogous to the work on plant tissues. When brain blots are labeled with DNA probes, multiple copies of DNA can be localized by comparing the distribution of label on the print with the adjacent section, which is stained histochemically. Similarly, whole animal frozen sections have been blotted onto nylon membranes, which are then treated with appropriate antibodies for localizing specific proteins (Lipkin & Oldstone, 1986).

It seems certain that tissue printing onto transfer membranes and other substrates to study many aspects of the development and physiology of animal cells will follow. The technique is simple and powerful and is versatile enough to isolate certain cell types from those of their neighbors. It is especially useful for localizing both specific proteins and specific nucleic acids on the tissue prints.

II. Isolation of Type-1 Astrocytes by Tissue Print Dissociation[1]

Tissue print dissociation exploits the specific binding properties of nitrocellulose to isolate certain cell types for electrophysiological or immunocytochemical study. The procedure results in glial cells that retain processes that would be lost during standard cell isolation techniques. This is useful because certain ion channel types are localized to processes.

Postnatal rat optic nerves contain three types of glial cells that can be distinguished by their antigenic properties. Cell separation by enzymatic digestion followed by printing onto nitrocellulose results in adherence of cells to the membrane. Some of the glial cells isolated in this way (Fig. 9.1) maintain their processes and present an exposed membrane surface that can be exploited for patch clamp studies of specific ion channels or for immunohistochemical localization of specific proteins.

A. Materials

1. Nitrocellulose membranes, 0.45–μm pore size (Schleicher & Schuell)
2. Nitrocellulose in amyl acetate, 1% solution (Ladd or Ted Pella)
3. Glass coverslips (13 mm) precleaned with 70% ethanol for several hours, rinsed in distilled water, and dried
4. Earle's balanced salts solutions (EBSS) containing calcium and magnesium (pH 7.4)
5. Vibratome for preparing sections of older tissue
6. Papain solution in Hepes buffer, 30 U/ml (Worthington) (see Barres, Koroshetz, Chung, & Corey, 1990)

[1]This section was adapted from "Ion Channel Expression by White Matter Glia: The Type-1 Astrocyte" B. A. Barres, W. J. Koroshetz, L. L. Y. Chung, and D. P. Corey. (1990). *Neuron* **5,** 527–544.

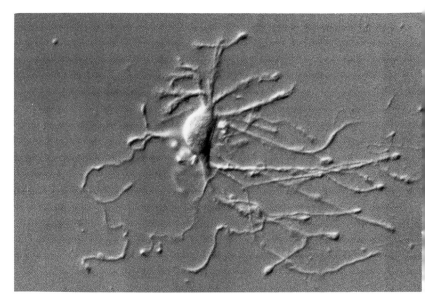

Figure 9.1 Optic nerve type-1 astrocyte isolated on a nitrocellulose-coated glass coverslip. From "Ion Channel Expression by White Matter Glia: The Type-1 Astrocyte" by B. A. Barres, W. J. Koroshetz, L. L. Y. Chung, and D. P. Corey (1990). *Neuron* **5,** 527–544.

7. Ovomucoid solution, 2 mg/ml (Calbiochem-Behring)
8. Normal saline solution: 5 mM Hepes buffer, 100 mM calcium, and 5 mM potassium

B. Procedure

1. Dissect optic nerves from postnatal Lang-Evans rats, and place in EBSS containing calcium and magnesium in a 35-mm petri dish at room temperature. Vibratome sections 50–100 mm thick of older adult tissue are similarly prepared.
2. Remove the EBSS solution and replace with papain solution; incubate for 10 min. Details of this procedure have been described by Huettner and Baughman (1986).
3. Remove the leptomeninges and incubate in papain for an additional 20 min (P2 nerves) to 75 min (P12 nerves or adult vibratome sections).

4. Replace the papain solution with 3 ml of ovomucoid solution in L15 medium containing DNAse.
5. Prewet nitrocellulose membranes by soaking in EBSS buffer. Dip glass coverslips into 1% nitrocellulose in amyl acetate, and dry overnight. Handle the coverslips carefully to avoid tearing the nitrocellulose film.
6. Print the tissue in a 35-mm petri dish under a dissecting microscope in normal saline solution. Gently press the enzymatically treated optic nerve tissue against the nitrocellulose membrane or nitrocellulose-coated glass coverslip. To accomplish this, first stabilize the tissue on a piece of nitrocellulose membrane (with the side of the tissue to be blotted faceup), and then invert the membrane over the coverslip. Lift the tissue from the coverslip and discard it. Place the print in fixative or the recording bath solutions, depending on the experiment to be performed.

III. Detection of Multiple DNA Copies by Brain Blotting[2]

Brain blotting is technically simple and inexpensive, and it can be used to screen specific brain regions at low magnification for the presence of multiple DNA copies (cellular or viral). The well-established *in situ* hybridization technique can then be used to further localize the signal at high magnification if more detail is needed.

Brain blotting is performed by printing thick frozen sections onto nylon membranes immediately after cryotome sectioning. A standard Southern hybridization is then performed on the membrane. A "replica" of the printed section is obtained by keeping on the glass slide the next frozen section cut, which is then stained for conventional histopathological analysis; the nuclei are counted to estimate the

[2]This section was contributed by Rafael Pont-Lezica and adapted from "Brain Blotting: A Method to Detect Multiple DNA Copies in Specific Brain Regions" by M. Hernandez Bronchud, S. Webb, and M. M. Esiri. (1988) **36**, pp. 1191-1195.

total amount of DNA in each section. This method of DNA detection, conveniently modified, might also be used to detect RNAs in specific coronal sections of whole brain before localization at high magnification by standard *in situ* hybridization techniques.

A. Materials

1. Standard 3-mm Whatman blotting paper and nylon membranes
2. 6 X SSC buffer: 0.9 M NaCl and 90 mM sodium citrate
3. 0.5 M NaOH
4. 1 M Tris-HCl (pH 7.5)
5. 0.1% SDS
6. 2 X SSC buffer
7. Prehybridization buffer: 50% formamide, 3 X SSC (0.45 M NaCl and 45 mM sodium citrate), 10 X Denhardt's solution (0.02% bovine serum albumin, 0.02% polyvinylpyrrolidone, 0.02% Ficoll, and 20 mg/liter heat-denatured, sonicated salmon sperm DNA)
8. Appropriate ^{32}P-DNA probe prepared by nick translation

B. Brain Blotting

1. Cut thick frozen sections (10–80 μm) of brain, frozen at −70°C without snap-freezing, with a standard cryotome, and immediately print onto 3-mm Whatman blotting paper by attaching the section to a glass slide and quickly pressing the glass slide on the blotting paper for ≈30 sec. It is important to lay the frozen section on the paper within 30 sec after cutting; otherwise, it remains permanently attached to the glass slide.
2. Gently remove the glass slide, leaving the frozen section on the blotting paper.
3. Make a replica of the blotted section by keeping the next frozen section cut on the glass slide. Stain this section for conventional histopathological analysis. Count the cell nuclei under a microscope at 400× magnification to estimate total DNA per section (assuming an average genome size of 3.5–5 pg of DNA per cell).

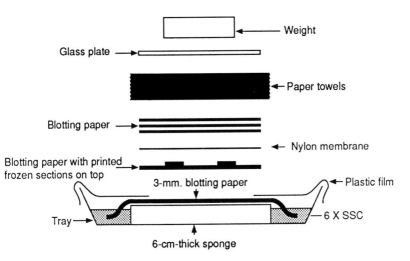

Figure 9.2 A Southern blotting apparatus is used for brain blotting. The printed frozen section is sandwiched between the blotting paper and the nylon membrane. From "Brain Blotting: A Method to Detect Multiple DNA Copies in Specific Brain Regions" by M. Hernandez Bronchud, S. Webb, and M. M. Esiri (1988). *Journal of Histochemistry and Cytochemistry* **36**, 1191–1195.

4. Place a nylon membrane of adequate size on the blotting paper overlying the frozen sections.
5. Transfer DNA to the nylon membrane by capillary diffusion, according to the conventional Southern technique (Southern, 1975), in 6 X SSC buffer for 24 hr, as shown in Fig. 9.2.
6. Rinse the membrane in 6 X SSC buffer, wrap it in plastic film, and fix the transferred DNA by exposing it to ultraviolet light (254 nm) for 3 min.
7. Denature the DNA by placing the membrane (DNA side up) on 3-mm blotting paper soaked in 0.5 M NaOH.
8. Neutralize by placing the membrane on blotting paper soaked in 1 M Tris-HCl.
9. Wash the membrane two times in a buffer solution of 0.1% SDS and 2 X SSC, and then wash it in 2 X SSC.

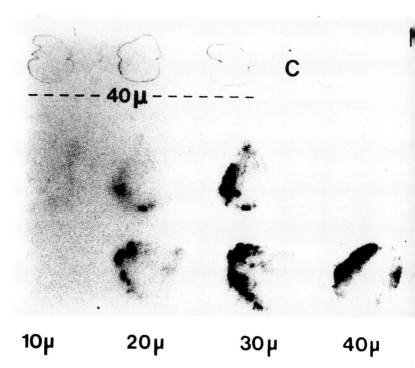

Figure 9.3 Three consecutive pencil-outlined adjacent sections of rat cerebellum (c) 40 μm thick are blotted onto the same nylon membrane as six consecutive frozen sections of human cerebellum of different thicknesses hybridized with nick-translated Alu probe [Alu sequence from BK(1.9)51, which hybridizes with total genomic human DNA, including both neuronal and glial] in high-stringency conditions to allow binding of probe to human Alu sequences but not to rat Alu sequences. Autoradiography exposure time is 3 days. The bar represents 1 cm. From "Brain Blotting: A Method to Detect Multiple DNA Copies in Specific Brain Regions" by M. Hernandez Bronchud, S. Webb, and M. M. Esiri (1988). *Journal of Histochemistry and Cytochemistry* **36**, 1191–1195.

C. Hybridization

1. Prehybridize the membrane at 42°C for 24 hr in a sealed polyethylene bag.
2. Hybridize in prehybridization buffer with the desired ^{32}P-DNA probe (2×10^7 to 6×10^7 cpm/mg of DNA), 10 mg/liter polyadenylic acid, and 5% dextran sulfate, either at 42°C with 50% formamide or at 68°C without formamide.
3. Wash the membrane two times in 2 X SSC at room temperature and two times (30 min each) in 0.1% SDS and 2 X SSC at 65°C.
4. After drying the membrane, autoradiograph with intensifying screens at −70°C for up to 10 days (see Fig. 9.3).

V. Localization of Proteins in Whole Animal Sections[3]

Laboratory animals are invaluable in the study of pathogenesis and management of infectious diseases. Nucleic acid techniques and immunotechniques have facilitated analysis of animal models of infectious diseases at the level of DNA, mRNA, and proteins. Whole animal sectioning has made possible the study of animal models at the molecular level in an anatomical context (Dubensky, Murphy, & Villareal, 1984; Southern, Blount, & Oldstone, 1984; Blount et al., 1986; Lipkin & Oldstone, 1986). It has been used to map the distribution of viruses and viral variants, chart the course of natural infections, follow the course of infections after experimental intervention, and examine the effects of infection on expression of host genes and gene products.

The original method involves collecting the frozen section on adhesive tape and immediately blotting the section onto nylon membrane for immunodetection of proteins (Figs. 9.4 and 9.5). For DNA

---

[3]This section was contributed by Rafael Pont-Lezica and adapted from "Whole Animal Section in situ Hybridization and Protein Blotting: New Tools in Molecular Analysis of Animal Models for Human Disease" by W.I. Lipkin, L.P. Villareal, and M.B.A. Oldstone (1989). Current Topics in Microbiology and Immunology **143**, pp. 33-54.

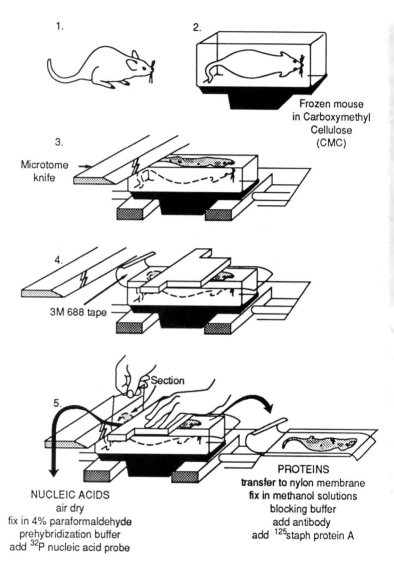

1.

2.

Frozen mouse
in Carboxymethyl
Cellulose
(CMC)

3.

Microtome
knife

4.

3M 688 tape

5.

Section

NUCLEIC ACIDS
air dry
fix in 4% paraformaldehyde
prehybridization buffer
add ^{32}P nucleic acid probe

PROTEINS
transfer to nylon membrane
fix in methanol solutions
blocking buffer
add antibody
add ^{125}staph protein A

Figure 9.4 A mouse is prepared for cryomicrotoming, *in situ* hybridization with whole animal sections, and protein blotting. (1, 2) The mouse is sacrificed and frozen in a block of carboxymethyl cellulose mounting medium. (3–5) 30-μm sagittal sections are cut with a cryomicrotome, picked up on tape, and processed either for *in situ* hybridization with probes of viral or host nucleic acids or for protein detection with antibodies to viral or host antigens. From "Whole Animal Section in situ Hybridization and Protein Blotting: New Tools in Molecular Analysis of Animal Models for Human Disease" by W. I. Lipkin, L. P. Villareal, and M. B. A. Oldstone 1989, *Current Topics in Microbiology and Immunology* **143,** pp. 33–54.

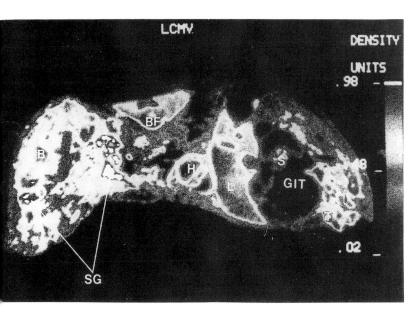

Figure 9.5 Whole animal section shows quantitative distribution of lymphocytic choriomeningitis virus (LCMV) antigens in a mouse persistently infected with LCMV. A sagittal mouse section was incubated with guinea pig antibody to LCMV and ^{125}I-labeled staphylococcal protein A. The autoradiograph was scanned with a laser densitometer for color-coded densitometry. Abbreviations: B, brain; BF, brown fat; GIT, gastrointestinal tract; H, heart; L, liver; S, spleen; SG, salivary glands; T, testes. LCMV antigen concentration is indicated by color: White is highest with decreasing concentrations shown in red, yellow, green, and blue (color bar at right of figure). From "Whole Animal Section in situ Hybridization and Protein Blotting: New Tools in Molecular Analysis of Animal Models for Human Disease" by W. I. Lipkin, L. P. Villareal, and M. B. A. Oldstone (1989). *Current Topics in Microbiology and Immunology* **143,** 33–54.

localization the sections are allowed to dry and then are fixed, and conventional *in situ* hybridization is then performed. The *in situ* hybridization procedure for whole animal sectioning is not included in this chapter. However, the procedure for small organs in animal tissues is described in Section III, and membranes blotted in the same way as for protein immunodetection can be used for RNA or DNA localization (see Chapter 7).

A. Materials

1. Bath of dry ice in ethanol
2. Cryomicrotome
3. Nylon membranes, 0.2-μm pore size (Pall)
4. 70% Laemmli buffer (Laemmli, 1970)
5. Blotto: 5% nonfat milk powder, 0.01% simethicone, and 0.001% thimerosal in phosphate buffer saline (20 mM phosphate buffer, pH 7.4, and 0.15 M NaCl)
6. Appropriate monoclonal or polyclonal antibody
7. ^{125}I-labeled staphylococcal protein A
8. 0.5 mM LiCl, 0.1 M Tris-HCl (pH 8.0), and 1% NP40
9. Kodak XRP-1 film or other fine-grain film

B. Whole Animal Sectioning

1. Anesthetize the animal for sacrifice by a deep axillar incision.
2. Freeze the shaved animal into a block of 3.5% carboxymethy cellulose in a bath of dry ice and ethanol.
3. Cut sections 20–40 μm thick with a cryomicrotome. Collec sections on transparent adhesive tape (3M No. 688) a shown in Fig. 9.4, and either transfer immediately to mem brane for protein blotting or allow to thaw and dry for 10–3 min at room temperature for *in situ* hybridization.

C. Protein Blotting

1. Place the section against a nylon membrane for 10–30 mi for passive protein transfer.
2. Immerse the membrane with the section attached in 70% Laemmli buffer and 30% methanol for 15 min.
3. Place the membrane in a solution of 10% acetic acid and 10% methanol (this dissolves the tape adhesive, allowing the tap to be stripped off), and fix the proteins to the membrane.

4. Rinse the membrane in water for 5–10 min. Strip tissue adhering to the membrane with a glass slide or a razor blade. Allow the membrane to dry, and either use immediately or store at 4°C.

5. Incubate the membrane in Blotto either overnight at 4°C or 2–4 hr at room temperature.

6. Incubate the membrane in the antibody diluted in Blotto for 4 hr at room temperature or overnight at 4°C. Polyclonal antibodies often react with gastrointestinal flora; incubating the antibodies with minced rodent intestines reduces this background reactivity.

7. Rinse the membrane in Blotto.

8. Incubate the membrane in ^{125}I-labeled staphylococcal protein A at a concentration of 5×10^5 cpm/ml for 1 hr.

9. Wash the membranes first in Blotto for 30 min, then in 0.5 mM LiCl, 0.1 M Tris, and 1% Nonidet P-40 for 20 min.

10. Rinse the membrane in water, drain, and set directly against the fine-grain film for 24–72 hr.

CHAPTER

Miscellaneous Applications of Tissue Printing

Rafael F. Pont-Lezica

Centre de Physiologie Végétale
Université Paul Sabatier
Toulouse, France

I. Overview

This book has presented a number of different uses for tissue printing, the most conspicuous being physical prints and chemical prints involving mainly macromolecules: proteins, glycoproteins, and nucleic acids. However, there are other applications of tissue printing, and surely many others will be developed in the future. This chapter presents some of the miscellaneous applications that involve macromolecules

Additional contributions to this chapter have been made by Cenk Suphioglu, Mohan B. Singh, and R. Bruce Knox, and Rosannah Taylor and Joseph Varner.

and small metabolites.

The immunocytochemical study of flowering plant pollen was developed to detect allergenic glycoproteins in certain pollen types (Hagman, 1964; Knox, Heslop-Harrison, & Reid, 1970). The technique was improved to detect minimal amounts of antigens from a small number of pollen grains; the improved technique was called "pollen printing" and used agarose film (Howlett, Knox, & Heslop-Harrison, 1973). Nitrocellulose membranes have now replaced the agarose film in pollen printing (O'Neil, Singh, & Knox, 1986, 1990). The principle of pollen printing is that pollen captured on a nitrocellulose filter will release allergenic proteins when air is passed through it if the filter is moistened with a buffer. The released proteins bind the nitrocellulose membrane with little diffusion. Mono- or polyclonal antibodies raised against the allergenic glycoproteins are used to detect the antigens, and enzyme-conjugated or fluorescent-labeled secondary antibodies are used to visualize them.

Some plant proteins have a relatively high content of cysteine residues (6–15%). This is the case for the cell wall thionins (Bohlmann et al., 1988), the cereal and solanaceous lectins (Vasta & Pont-Lezica, 1990), and the phloem proteins known as P proteins. These cysteine-rich proteins can be detected and localized by using specific antibodies or by taking advantage of their high content of cysteinyl residues. S-Carboxymethylation of cysteinyl residues with a fluorescent reagent is a rapid method for visualizing these proteins. Derivatization of cysteine residues, with or without previous reduction of disulfide bonds, can be accomplished on tissue prints (Pont-Lezica & Varner, 1989b).

Small metabolites are also blotted onto the membranes with little diffusion. Hydrophobic molecules are retained by nitrocellulose or other hydrophobic membranes, and they can be visualized with appropriate probes. An example of a small metabolite is presented in Section V, in which naturally fluorescent compounds are localized. Taylor and Varner have used tissue printing not only for the localization within the tissue but also for a chromatographic separation of some of the components, showing the presence of at least two classes of fluorescent compounds in the secretory ducts of celery petioles. Tissue printing methods for rapid plant screening in the search for interesting natural products may develop rapidly in the near future. Another example of localizing small metabolites is in Section IV, in which ascorbic acid is localized by the acidic silver nitrate method. Ascorbic acid is one of the few compounds that reduce silver in acidic conditions.

The main problem with localizing small metabolites is that they can be washed out by the solutions used for development. However, different techniques, such as the previous imbibition of the membrane with the appropriate reagents or the use of an overlay gel containing the reagents, can prevent this. In these cases, little diffusion is expected.

II. Detection of Airborne Grass Pollen Allergens[1]

Certain types of wind-borne pollen, such as grass pollen, contain allergenic glycoproteins. The search for these allergens led to the first immunocytochemical studies of flowering plant antigens carried out on pollen (Hagman, 1964; Knox, Heslop-Harrison, & Reed, 1970). Immunofluorescent studies of ragweed pollen, freeze-sectioned within the anther to minimize diffusion of antigens, showed that the major ragweed allergen, antigen E, is concentrated in the pollen walls and surfaces (Knox & Heslop-Harrison, 1971). Similar techniques using polyclonal antibodies determined the surface location of grass pollen allergens (Howlett, Vithanage, & Knox, 1981).

Pollen from many plants, e.g., the grasses, cannot be identified morphologically at the genus level with either light microscopy or scanning electron microscopy. Moreover, scanning electron microscopy is impractical for routine high-turnover sampling. Therefore, a new method for microscopic identification of pollen type would be useful, particularly for genera producing morphologically similar pollen simultaneously in the same geographic area.

[1]This section was contributed by Cenk Suphioglu, Mohan B. Singh, and R. Bruce Knox. This work was supported in part by the National Health and Medical Research Council, the Australian Research Council, and the University of Melbourne Postgraduate Scholarship to Cenk Suphioglu.

Biochemical analyses of these allergens require large quantities of pollen; more than 100 mg of pollen is sometimes required. The pollen print technique permits extraction and visualization of microquantities of protein from the pollen walls (Howlett, Knox, & Heslop-Harrison, 1973). The method involves applying dry pollen grains, attached to a piece of adhesive tape, to an agarose film on a microscope slide. The pollen protein components are released into the moist agarose, the pollen grains are removed with the tape, and the agarose film is dried and stained. Immunofluorescence studies of the agarose films with polyclonal antibodies have shown the progressive release of antigen E from the surface and wall pores of ragweed pollen (Howlett, Knox, & Heslop-Harrison, 1973).

A modification of the pollen print method, in which moist nitro-cellulose or cellulose nitrate paper, rather than agarose, is used as the substrate to capture pollen proteins, has been previously described (O'Neil, Singh, & Knox, 1986, 1990). Nitrocellulose membrane is used extensively for electroblotting protein from polyacrylamide gels (Towbin, Staehlin, & Gordon, 1986) and, because proteins bind electrostatically to the membrane, is also used for the dot blotting of antigens for immunoassay studies (Hawkes, Niday, & Gordon, 1982; Singh & Knox, 1985). Immunofluorescence microscopy of nitrocellulose membranes has also been used to identify cross reactivity among various grass pollens and polyclonal antibodies to Bermuda grass pollen (Schumacher, Griffith, & O'Rourke, 1988).

We reported the first tissue printing on nitrocellulose membranes in 1986 (O'Neil, Singh, & Knox, 1986). Here, we present a new application of this immunochemical method, in which pollen grains of *Lolium perenne* (ryegrass) are distinguished from morphologically identical grass pollen types present in the atmosphere; it is based on the known immunologic uniqueness of ryegrass allergens (Fig. 10.1). The method is practical and efficient and has many other advantages; for example, after the diffusion blotting step, sampled nitrocellulose filters can be stored dry in a petri dish for a long time before immunodetection is undertaken.

This protocol describes the membrane print technique with a series of simple instructions. Air is passed through a dry nitrocellulose filter to capture airborne pollen. Later, allergenic proteins are allowed

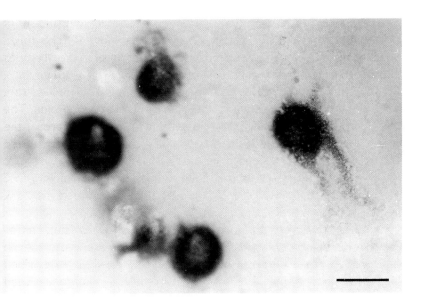

Figure 10.1 Pollen protein prints of ryegrass, Lolium perenne, collected on a nitrocellulose filter from the atmosphere, show anti-LolpI monoclonal antibody binding with the peroxidase indirect labeling method. The bar represents 50 mm. Quantitative densitometric analysis of these nitrocellulose pollen prints could be used to determine the relative concentration of aeroallergens in relation to the prevalence of the plants in the region sampled. Thus, this method may potentially standardize the assay for daily measurements of airborne allergens during the hayfever seasons. Similarly, because different grass pollens are difficult to identify morphologically, this method could be used, with appropriate specific monoclonal antibodies, as an immunodetection assay for identifying the different pollen types.

to leach out of the captured pollen by diffusion blotting on a moistened Whatman paper moistened with phosphate-buffered saline (PBS) and to bind to the nitrocellulose filter.

A. Materials

1. Sartorius MD 8 air sampler with 80-mm nitrocellulose filter with 0.8-μm pore size, or Burkhard 7-day volumetric spore trap.

2. PBS.
3. Calberla's stain: 17% glycerin, 32% ethyl alcohol, 2–3 drops saturated aqueous basic fuchsin, 2–3 drops melted glycerin jelly in distilled water.
4. Blotto: 10% w/v skim milk powder in PBS.
5. Tween PBS (TPBS): 0.1% Tween-20 in PBS.
6. 0.5% bovine serum albumin (BSA).
7. Mono- or polyclonal antibodies against grass-pollen-allergic subject.
8. Horseradish-peroxidase-conjugated sheep anti-mouse immunoglobulins (Silenus) diluted 1:1000 in PBS with 0.5% BSA. If rabbit serum is used instead of monoclonal antibodies, use immunoglobulins to human IgE (DAKO), diluted 1:200 in PBS with 0.5% BSA, and horseradish-peroxidase-conjugated anti-rabbit IgG (Promega), diluted 1:2500 in PBS with 0.5% BSA.
9. Peroxidase substrate: 0.06% 4-chloro-1-naphthol, 0.006% H_2O_2, and 20% methanol in PBS.

B. Pollen Sampling

In our experiments, we sampled air for 2 hr in the mornings during peak hay fever season, late spring to summer, in northerly winds, using a Sartorius MD 8 air sampler operating at 4.5 m^3/h. Alternatively, air may be sampled through a filter unit connected in series with a vacuum pump operating under conditions similar to those just described (Habenicht, Burge, Muilenberg, & Solomon, 1984).

Another successful method for pollen sampling uses a Burkard spore trap. The trap is calibrated for air sampling at a set flow rate, and the pollen is trapped on a Burkard tape coated with a nonaqueous Dow Corning adhesive and enclosed in the collecting assembly of the trap. After collection, the Burkard tapes are removed and saturated with PBS. Then a PBS-moistened nitrocellulose membrane is applied to the surface of the tape, and 10 min is allowed for protein transfer. The nitrocellulose membrane is then treated as described in this protocol (Schumacher, Griffith, & O'Rourke, 1988).

C. Procedure

1. After sampling, carefully remove the nitrocellulose filter from the unit with forceps and place faceup on a PBS-moistened Whatman paper; enclose in a petri dish for 1 hr. This facilitates the binding of readily leachable allergenic proteins of the pollen by diffusion blotting. Similarly, if a mesh or Millipore filter is used for the air sampling, it should be placed faceup on a PBS-moistened Whatman paper, with a moistened nitrocellulose membrane placed on top for 1 hr, to generate a replica print of the captured pollen proteins.

2. Cut small portions of the nitrocellulose filter, and mount them in Calberla's stain (to stain the pollen red) on a microscope slide. Gently apply a coverslip, and observe with a light microscope to confirm the presence of pollen grains.

3. If pollen grains are present, block the free protein binding sites of the nitrocellulose filter by washing in Blotto for 1–2 hr with gentle shaking. Pollen grains will be shaken from the nitrocellulose filters into the Blotto. This pollen grain detachment poses no problems because the readily leachable allergens from the pollen grains have already bound to the nitrocellulose filter by diffusion blotting. Be sure to make a negative control by treating a clean nitrocellulose filter section in parallel with those used for the sampling. Optionally, make a positive control by placing a known pollen on the nitrocellulose filter.

4. Wash the nitrocellulose filter once with TPBS and then two times with PBS only (5 min each time) with gentle shaking.

5. Incubate the nitrocellulose filter in the mono- or polyclonal antibody (or the serum of a grass-pollen-allergic subject diluted 1:4) in PBS with 0.5% BSA for 3 hr on a gently rotating wheel.

6. After TPBS and PBS washing (as in step 4), incubate the nitrocellulose filters in horseradish-peroxidase-conjugated sheep anti-mouse immunoglobulins for 1 hr. If the nitrocellulose filters were incubated in rabbit serum, after TPBS and PBS washing they should be incubated first in rabbit

immunoglobulins to human IgE and then in horseradish-peroxidase-conjugated anti-rabbit IgG for 1 hr each on a gently
rotating wheel, with a TPBS and PBS washing between the
two antibody incubations.

7. Once again, wash in TPBS and PBS, and develop the nitrocellulose filters in freshly prepared peroxidase substrate at 37°C
in the dark for 5–10 min (≤30 min), and stop the development by gently but thoroughly washing in distilled water.

8. Mount small sections of the developed and control nitrocellulose filters in water on a slide, apply a coverslip, and
observe the pollen print by bright-field microscopy using
transmitted illumination. Photograph with Tmax 100 or any
other compatible film (Fig. 10.1).

Note

A black (instead of white) nitrocellulose filter can be used in the
sampling procedure, and the pollen allergens can be identified by antibody labeled with fluorescein isothiocyanate (FITC) (instead of the
horseradish peroxidase-conjugated secondary antibody) and the nitrocellulose filter can then be observed under a fluorescent microscope.
Although one can achieve good results by using the FITC-labeled secondary antibody, the fluorescent dye tends to fade away relatively
quickly. This makes the procedure impractical if many nitrocellulose filters need to be processed and observed simultaneously.

III. Localization of Cysteine-Rich Proteins

Cysteine-rich proteins in plants have been localized by immunocytochemical techniques (Bohlmann et al., 1988; Mishkind, Raikhel, Palevitz
& Keegstra, 1982). This protocol presents a rapid method for the tissue localization of cysteine-rich proteins that uses tissue printing on
nitrocellulose membranes followed by reduction and S-carboxymethylation of cysteinyl residues with the fluorescent dye N-iodoacetyl- N′

-5-sulfo-1-naphthyl)ethylenediamine (5-I-AEDANS) (Gorman, 1987). S-Carboxymethylation without previous reduction allows the ulfhydryl groups of the cysteines to be detected. This method has een used to localize cysteine-rich proteins in soybean (Fig. 10.2), ger- ninating barley, and potato tuber tissues (Pont-Lezica & Varner, 989b).

A. Materials

1. Nitrocellulose (Schleicher & Schuell) or Immobilon P (Milli- pore) membrane, 0.45-μm pore size
2. Tween Tris buffer saline (TTBS): 0.05% Tween-20, 20 mM Tris-HCl (pH 7.4), and 0.5 M NaCl
3. Reducing agent: 20 mM dithiothreitol (DTT) or β-mercap- toethanol (β-ME)
4. 20 mM 5-I-AEDANS (Sigma) in 0.1 M NH_4HCO_3
5. 20 mM iodoacetamide in 0.1 M NH_4HCO_3
6. Manual longwave (360 μm) ultraviolet (UV) light

B. Procedure

1. Soak the nitrocellulose membrane in TTBS for 30 min, and dry on paper towels (nitrocellulose without previous treat- ment can also be used, but a better transfer can be obtained with pretreated membrane). If Immobilon P is used, wet the membrane with methanol for 3 sec, and then soak it in TTBS as described for nitrocellulose.
2. Cut the tissue to be printed with a new razor blade, remove the liquid on the surface of the freshly cut surface with Kimwipes, and press the section onto the membrane for 15–30 sec. The tissue print can be dried and kept in a refrig- erator, or it can be processed immediately for detection of cysteine residues.
3. Incubate the print with the reducing agent at 60°C for 2 hr in a petri dish.
4. Wash the membrane with water three times for 5 sec each.

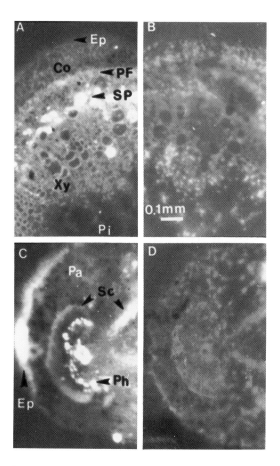

Figure 10.2 Sections of soybean stem (A, B) and the ventral suture of the pod (C, D) were either stained with 5-I-AEDANS (A, C) or treated with iodoacetamide before fluorescent dying (B, D). Abbreviations: Ep, epidermis; Co, cortex; PF, phloem fibers; SP, secondary phloem; Xy, xylem; Pi, pith; Pa, parenchyma; Sc, sclerenchyma; Ph, phloem. Note the circle of highly fluorescent dots associated with the secondary phloem of stem and pod from P proteins (phloem proteins) rich in free cysteine residues. From "Histochemical Localization of Cysteine-Rich Proteins by Tissue Printing on Nitrocellulose" by R. F. Pont-Lezica and J. E. Varner (1989). *Analytical Biochemistry* **182**, 334–337.

5. Incubate the membrane with 20 mM 5-I-AEDANS solution for 2 hr at room temperature. Parallel controls are incubated with 20 mM iodoacetamide in 0.1 M NH₄CO₃ for 2 hr and then with the fluorescent dye (5-I-AEDANS).

6. Wash the membrane several times with TTBS until the fluorescent background disappears.

7. When dry, observe the prints under UV light with a low-power microscope or with an epifluorescent microscope if higher magnification is desired. Use ISO 400 film to photograph the prints. Use a yellow filter (Tiffen No. 12) for black-and-white film or a Plexiglas filter for color film (to avoid UV irradiation of the film and maintain the original colors). If an epifluorescent microscope is used, the filters are already installed on the microscope. **Caution:** Ultraviolet radiation is dangerous, particularly to the eyes. To minimize exposure, make sure that the UV light source is shielded and wear protective goggles or a safety mask to block the UV light.

V. Localization of Ascorbic Acid[2]

In the late 1920s Szent-Gyorgi used acidic silver nitrate to determine the occurrence of and to follow the purification of "hexuronic acid," now known as ascorbic acid. We used the same method here to show the localization of acidic AgNO₃-reducing material presumed to be ascorbate (Fig. 10.3).

A. Materials

1. Nitrocellulose membrane, 0.45-μm pore size (Schleicher & Schuell)
2. 4% AgNO₃ in 100 mM sodium acetate buffer (pH 4.0)
3. Celery plants (*Apium graveolens*) cultivated in the greenhouse under natural daylight

[2]This section was contributed by Rosannah Taylor and Joseph E. Varner.

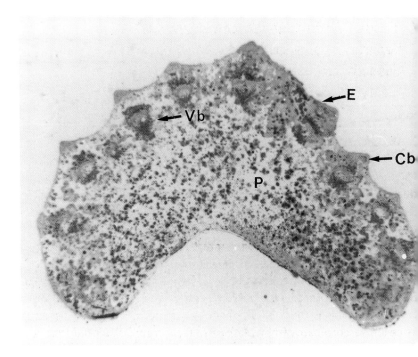

Figure 10.3 A cross section from a freshly cut celery stalk was printed c nitrocellulose membrane, and the membrane was soaked in 4% AgNO₃ in 100 m sodium acetate buffer (pH 4.0) to show the localization of ascorbic acid in the tissu The membrane was dried and photographed. Abbreviations: E, epidermis; C colleuchyma bundle, P, pith; VB, vascular bundle.

B. Procedure

1. Cut a 2–3 mm-thick cross section from the central portio of a celery petiole, and gently preblot the section on a sep rate piece of membrane or filter paper before printing. Plac the section on an untreated nitrocellulose membrane. Place piece of smooth paper over the membrane to protect it fro fingerprints, and then apply pressure for 10–30 sec.

2. Soak the printed membrane in the AgNO₃ solution. Usually pattern emerges in a few seconds (Fig. 10.3). Examine unde a microscope, and photograph.

Tissue Printing and Chromatographic Separation of Autofluorescent Substances in Secretory Ducts in Celery Petioles[3]

Celery secretory ducts contain many substances, some of which are partly responsible for the characteristic odor and flavor of celery (*Apium graveolens* L.). Secretory products are important commercial sources of many chemicals and medicinal compounds (Alfermann & Reinhard, 1987). Tissue printing can be used to rapidly screen plant tissues for these secretory products. When a freshly cut section of celery is printed on nitrocellulose and observed under longwave UV light, fluorescent patterns from secretory cells may be seen (Varner, Taylor, Cassab, Lin, Yuen & Pont-Lezica, 1989). We have used tissue printing to study soluble fluorescent compounds from secretory ducts in *Apium graveolens* L. This is the first report showing the use of tissue printing on cellulose chromatography plates for separating soluble fluorescent compounds (Fig. 10.4).

A. Materials

1. Cellulose thin-layer chromatography plates, 10×20 cm, MN 300 (Analtech)
2. Chloroform–methanol solution (9:1 v/v) as developer
3. Manual longwave UV light (366 nm) (Blak-Ray UVL-21, UVP, Inc.)
4. 35-mm camera
5. Kodak Tmax black-and-white print film, ISO 400
6. Celery plants (*Apium graveolens* L.) cultivated in the greenhouse under natural daylight

B. Procedure

1. Make a 2–3 mm-thick cross section of the central portion of a celery petiole. Gently preblot the section on a separate

---

[3]This section was contributed by Rosannah Taylor and Joseph E. Varner.

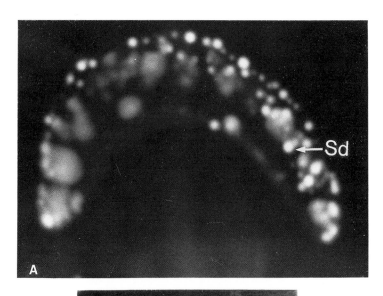

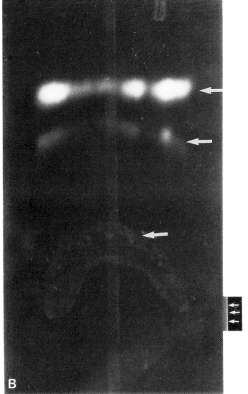

piece of membrane or filter paper before printing. Place the section on an untreated cellulose thin-layer chromatography plate (1–1.5 cm from the bottom of the plate). Place a piece of smooth paper over the section to protect the plate from fingerprints, and then apply pressure for 10–30 sec.

2. Develop the plate in a chromatography tank containing chloroform–methanol solution until the front has migrated 2–5 cm from the origin.

3. Dry the plate, and observe under UV illumination.

4. Photograph the plate with the Tmax film and a yellow filter (Tiffen No. 12). For color photographs use a Plexiglas filter to avoid UV irradiation of the film. **Caution:** Ultraviolet radiation is dangerous, particularly to the eyes. To minimize exposure, make sure that the UV light source is shielded and wear protective goggles or a safety mask to block the UV light.

Figure 10.4 (A) A tissue print of a celery petiole made on a microcrystalline cellulose chromatography sheet shows the secretory ducts (Sd). (B) Separation bands (arrows) result from chromatography of a tissue print of a celery petiole in chloroform–methanol (9:1 v/v).

References

Alfermann, A. W., & Reinhard, E. (1977, September). Production of natural compounds by cell culture methods. Symposium conducted at the University of Tubingen, Tubingen, Federal Republic of Germany.

Alpenfels, W. F. (1981). A rapid and sensitive method for the determination of monosaccharides as their dansyl hydrazones by high-performance liquid chromatography. *Analytical Biochemistry* **114,** 153–157.

Azanza-Corrales, R. (1991). The farmed *Eucheuma* species in Danajon Reef, Philippines: Vegetative and reproductive structure. *Journal of Applied Phycology* **2,** 57–62.

Azanza-Corrales, R., & Dawes, C. J. (1989). Wound healing in cultured *Eucheuma alvarezii* var. *tambalang* Doty. *Botanica Marina* **32,** 229–234.

Bailey, B. A., Dean, J. F. D., & Anderson, J. D. (1989). An ethylene biosynthesis-inducing endoxylanase elicits electrolyte leakage and necrosis in *Nicotiana tabacum* cv. Xanthi leaves. *Plant Physiology* **94,** 1848–1854.

Bailey, B. A., Dean, J. F. D., & Anderson, J. D. (1991). An ethylene biosynthesis-inducing endoxylanase is translocated through the xylem of *Nicotiana tabacum* cv. Xanthi plants. *Plant Physiology* **97,** 1181–1186.

Barres, B. A., Koroshetz, W. J., Chung, L. L. Y., & Corey, D. P. (1990). Ion channel expression by white matter glia: The type-1 astrocyte. *Neuron* **5,** 527–544.

Benhamou, N., & Oullette, G. B. (1986). Ultrastructural localization of glycoconjugates in the fungus *Ascocalyx abietina*, the Scleroderris canker agent of conifers, using lectin–gold complexes. *Journal of Histochemistry and Cytochemistry* **34,** 855–867.

Biles, C. L., Martyn, R. D., & Wilson, H. D. (1989). Isozyme and systemic resistance induced in watermelon by *Formae speciales* or *Fusarium oxysporum*. *Phytopathology* **94,** 856–860.

Blake, M. S., Johnston, K. H., Russell-Jones, G. J., & Gotschlich, E. C. (1981). A rapid, sensitive method for detection of alkaline phosphatase-conjugated anti-antibody on Western blots. *Analytical Biochemistry* **136,** 175–179.

Blount, P., Elder, J., Lipkin, W. I., Southern, P. J., Buchmeier, M. J., & Oldstone, M. B. A. (1986). Dissecting the molecular anatomy of the nervous system: Analysis of RNA and protein expression in whole body sections of laboratory animals. *Brain Research* **382**, 257–265.

Bohlmann, H., Clausen, S., Behnke, S., Giese, H., Hiller, C. C., Reimann-Philipp, C., Schrader, G., Barkholt, V., & Apel, K. (1988). Leaf-specific thionins of barley—A novel class of cell wall proteins toxic to plant-pathogenic fungi and possibly involved in the defence mechanism of plants. *EMBO Journal* **7**, 1559–1565.

Bozart, C., & Boyer J. S. (1987). Cell wall proteins at low water potentials. *Plant Physiology* **85**, 261–267.

Cassab, G. I. (1986). Arabinogalactan proteins during the development of soybean root nodules. *Planta* **168**, 441–446.

Cassab, G. I., Lin, J.-J., Lin, L.-S., & Varner, J. E. (1988). Ethylene effect on extensin and peroxidase distribution in the subapical region of pea epicotyls. *Plant Physiology* **88**, 522–524.

Cassab, G. I., Nieto-Sotelo, J., Cooper, J. B., Van Holst, G. J., & Varner, J. E (1985). A developmentally regulated hydroxyproline-rich glycoprotein from the cell walls of soybean seed coats. *Plant Physiology* **77**, 532–535.

Cassab, G. I., & Varner, J. E. (1987). Immunocytolocalization of extensin in developing soybean seed coats by immunogold–silver staining and by tissue printing on nitrocellulose paper. *Journal of Cell Biology* **105**, 2581–2588.

Cassab, G. I., & Varner, J. E. (1988). Cell wall proteins. *Annual Review of Plant Physiology and Plant Molecular Biology* **39**, 321–353.

Cassab, G. I., & Varner, J. E. (1989). Tissue printing on nitrocellulose paper: A new method for immunolocalization of proteins, localization of enzyme activities and anatomical analysis. *Cell Biology International Reports* **13**, 1147–1152.

Chen, J. A., & Varner, J. E. (1985a). Isolation and characterization of cDNA clones for carrot extensin and proline-rich 33 kDa protein. *Proceedings of the National Academy of Sciences of the United States of America* **82**, 4399–4403.

Chen, J. A., & Varner, J. E. (1985b). An extracellular matrix protein in plants: Characterization of a genomic clone for carrot extensin. *EMBO Journal* **4**, 2145–2151.

Condit, C., & Meagher, R. B. (1987). Expression of a gene encoding a glycine-rich protein in petunia. *Molecular and Cellular Biology* **7**, 4273–4279.

Cox, K. H., & Goldberg, R. B. (1988). Analysis of plant gene expression. In C. H. Shaw (Ed.), *Plant molecular biology: A practical approach* (pp. 1–34). Oxford, England: IRL Press.

Daoust, R. (1957). Localization of deoxyribonuclease in tissue sections. A new approach to the histochemistry of enzymes. *Experimental Cell Research* **12**, 203–211.

Daoust, R. (1965). Histochemical localization of enzyme activities by substrate film methods: Ribonucleases, deoxyribonucleases, proteases, amylase and hyaluronidase. *International Review of Cytology* **18**, 191–221.

Datta, K., Schmidt, A., & Marcus, A. (1989). Characterization of two soybean repetitive proline-rich proteins and a cognate cDNA from germinated axes. *Plant Cell* **1**, 945–952.

Dean J. F., Gamble, H. R., & Anderson, J. D. (1989). The ethylene biosynthesis-inducing xylanase: Its induction in *Trichoderma viride* and certain plant pathogens. *Phytopathology* **79**, 1071–1078.

del Campillo, E., Reid, P. D., Sexton, R., & Lewis, L. N. (1990). Occurrence and localization of 9.5 cellulase in abscising and nonabscising tissues. *Plant Cell* **2**, 245–254.

Della-Penna, D., Christoffersen, R. E., & Bennett, A. B. (1986). Biotinylated proteins as molecular weight standards on Western blots. *Analytical Biochemistry* **52**, 329–332.

Dimitriadis, G. J. (1979). Effect of detergents on antibody–antigen interactions. *Analytical Biochemistry* **98**, 445–451.

Doty, M. S. (1985). *Eucheuma alvarezii*, sp. nov. (Gigartinales, Rhodophyta) from Malaysia. In I. A. Abbott & J. H. Norris (Eds.), Taxonomy of economic seaweeds: With reference to some Pacific and Caribbean species (pp. 37–45). La Jolla, CA: California Sea Grant College Program.

Doty, M. S. (1986). The production and uses of *Eucheuma*. In M. S. Doty, J. F. Caddy, & B. Santelices (Eds.), *Case studies of seven commercial seaweed resources: FAO fisheries technical papers, no. 281* (pp. 123–164). Lanham, MD: UNIPUB.

Dubensky, T. W., Murphy, F. A., & Villareal, L. P. (1984). The detection of DNA and RNA virus genomes in the organ system of whole mice: Patterns of mouse organ infection by polyoma virus. *Journal of Virology* **50**, 779–783.

Eckhardt, A. E., Hayes, J. E., & Goldstein, J. J. (1976). A sensitive fluorescent method for the detection of glycoproteins. *Analytical Biochemistry* **73**, 192–197.

Eisner, T., Eisner, M., & Meinwald, J. (1987). Technique for visualization of epidermal glandular structures in plants. *Journal of Chemical Ecology* **13**, 943–946.

Esau, K. (1965). *Plant Anatomy*. New York: John Wiley.

Estep, T. N., & Miller, T. J. (1986). Optimization of erythrocyte membrane glycoprotein fluorescent labeling with dansyl hydrazine after polyacrylamide gel electrophoresis. *Analytical Biochemistry* **157**, 100–105.

Feigl, F. (1966). *Spot tests in organic analysis* (7th English ed., p. 132). Amsterdam: Elsevier.

Fincher, G. B., Stone, B. A., & Clarke, A. E. (1983). Arabinogalactan proteins: Structure, biosynthesis, and function. *Annual Review of Plant Physiology* **34**, 47–70.

Flurkey W. H., & Ingebrigtsen, J. (1989). Polyphenol oxidase activity and enzymatic browning in mushrooms. In J. J. Jen (Ed.), *Quality factors of fruit and vegetables: Chemistry and technology* (pp. 44–54). Washington, DC: American Chemical Society.

Franco, A. R., Gee, M. A., & Guilfoyle, T. J. (1990). Induction and superinduction of auxin-responsive mRNAs with auxin and protein synthesis inhibitors. *Journal of Biological Chemistry* **265**, 15845–15849.

Franssen, H. J., Nap, J. P., Gloudemans, T., Stiekema, W., & Van Dam, H. (1987). Characterization of cDNA for nodulin-75 of soybean: A gene product involved in early stages of root development. *Proceedings of the National Academy of Sciences of the United States of America* **84**, 4495–4499.

Fritz, S. E., Hood, K. R., & Hood, E. E. (1991). Localization of soluble and insoluble fractions of hydroxyproline-rich glycoprotein in maize pericarp. *Journal of Cell Science* **98**, 545–550.

Gabius, S., Hellmann, K.-P., Hellmann, T., Brinck, U., & Gabius, H.-J. (1989). Neoglycoenzymes: A versatile tool for lectin detection in solid-phase assays and glycohistochemistry. *Analytical Biochemistry* **182**, 447–451.

Gaddum, P., & Blandau, R. J. (1970). Proteolytic reaction of mammalian spermatozoa on gelatin membranes. *Science* **170**, 749–751.

Gahan, P. B. (1984). *Plant histochemistry and cytochemistry.* New York: Academic Press.

Glass, W. F., II, Briggs, R. C., & Hnilica, L. S. (1981). Use of lectins for detection of electrophoretically separated glycoproteins transferred onto nitrocellulose sheets. *Analytical Biochemistry* **115**, 219–224.

Goding, J. W. (1983). *Monoclonal antibodies: Principles and practice.* New York: Academic Press.

Gorman, J. J. (1987). Fluorescent labeling of cysteinyl residues to facilitate electrophoretic isolation of proteins for amino-terminal sequence analysis. *Analytical Biochemistry* **160**, 376–387.

Gretz, M., Wu, Y., Vreeland, V., & Scott, J. (1990). Iota-carrageenan biogenesis in the red alga *Agardhiella subulata* is Golgi mediated. *Journal of Phycolology* **26**(Suppl.), 14.

Habenicht, H. A., Burge, H. A., Muilenberg, M. L., & Solomon, W. R. (1984). Allergen carriage by atmospheric aerosol. II. Ragweed-pollen determinants in submicronic atmospheric fractions. *Journal of Allergy and Clinical Immunology* **74**, 64–67.

Haberlandt, G. (1914). *Physiological plant anatomy*. New Delhi: Today and Tomorrow's Book Agency. Reprinted.

Hagman, M. (1964). The use of disc electrophoresis and serological reactions in the study of pollen and style relationships. In H. F. Linskens (Ed.), *Pollen: Physiology and fertilisation* (pp. 244–250). Amsterdam: North Holland.

Hancock, K., & Tsang, V. C. W. (1983). India ink staining of protein on nitrocellulose paper. *Analytical Biochemistry* **133**, 157–162.

Harris, N., & Chrispeels, M. J. (1975). Histochemical and biochemical observations on storage protein metabolism and protein body autolysis in cotyledons of germinating beans. *Plant Physiology* **56**, 292–299.

Hawkes, R., Niday, E., & Gordon, J. (1982). A dot-immunobinding assay for monoclonal and other antibodies. *Analytical Biochemistry* **119**, 142–147.

Hay, G. W., Lewis, B. A., & Smith, F. (1965). Periodate oxidation of polysaccharides: General procedures. *Methods of Carbohydrate Chemistry* **5**, 357–377.

Hayashi, Y., & Ueda, K. (1987). Localization of mannose, N-acetylglucosamine and galactose in the Golgi apparatus, plasma membranes and cell walls of *Scenedesmus acuminatus*. *Plant Cell Physiology* **28**, 1357–1362.

Hernandez Bronchud, M., Webb, S., & Esiri, M. M. (1988). Brain blotting: A method to detect multiple DNA copies in specific brain regions. *Journal of Histochemistry and Cytochemistry* **36**, 1191–1195.

Hoggart, R. M., & Clarke, A. E. (1984). Arabinogalactans are common components of angiosperm styles. *Phytochemistry* **23**, 1571–1573.

Holt, C. A., & Beachy, R. N. (1991). *In vivo* complementation of infectious transcripts from mutant tobacco mosaic virus cDNAs in transgenic plants. *Virology* **181**, 109–117.

Hong, J. C., Nagao, R. T., & Key, J. L. (1989). Developmentally regulated expression of soybean proline-rich cell wall protein genes. *Plant Cell* **1**, 937–943.

Hood, E. E., Shen, Q. X., & Varner, J. E. (1988). A developmentally regulated hydroxyproline-rich glycoprotein in maize pericarp cell walls. *Plant Physiology* **87**, 138–142.

Hood, K. R., Baasiri, R. A., Fritz, S. E., & Hood, E. E. (1991). Biochemical and tissue print analyses of hydroxyproline-rich glycoproteins in cell walls of sporophytic maize tissues. *Plant Physiology* **96**, 1214–1219.

Howlett, B. J., Knox, R. B., & Heslop-Harrison, J. (1973). Pollen wall proteins: Release of the allergen Antigen E from intine and exine sites in pollen grains of ragweed and *Cosmos*. *Journal of Cell Science* **13**, 603–619.

Howlett, B. J., Vithanage, H. I. M. V., & Knox, R. B. (1981). Immunofluorescence localization of two water soluble glycoproteins, including the major allergen of rye-grass, *Lolium perenne. Histochemistry Journal* **13**, 461–480.

Hsu, H. T., & Lawson, R. H. (in press). Direct tissue blotting for detection of tomato spotted wilt virus in *Impatiens. Plant Disease.*

Huettner, J. E., & Baughman, R. W. (1986). Primary culture of identified neurons from the visual cortex of postictal rats. *Journal of Neuroscience* **6**, 3044–3060.

Jacobsen, J. V., & Knox, R. B. (1973). Cytochemical localization and antigenicity of α-amylase in barley aleurone tissue. *Planta* **112**, 213–224.

Keller, B., & Lamb, C. J. (1989). Specific expression of a novel cell hydroxyproline-rich glycoprotein gene in lateral root initiation. *Genes and Development* **3**, 1639–1646.

Keller, B., Sauer, N., & Lamb, C. J. (1988). Glycin-rich cell wall proteins in bean: Gene structure and association of the protein with the vascular system. *EMBO Journal* **7**, 3625–3633.

Keller, B., Templeton, M. D., & Lamb, C. J. (1989). Specific localization of a plant cell wall glycine-rich protein in protoxylem cells of the vascular system. *Proceedings of the National Academy of Sciences of the United States of America* **86**, 1529–1533.

Kieliszewski, M., & Lamport, D. T. A. (1987). Purification and partial characterization of a hydroxyproline-rich glycoprotein in a graminaceous monocot, Zea mays. *Plant Physiology* **85**, 823–827.

Kieliszewski, M., Leykam, J. F., & Lamport, D. T. A. (1990). Structure of the threonine-rich extensin from *Zea mays. Plant Physiology* **92**, 316–326.

Knox, J. P. (1990). Emerging patterns of organization at the plant cell surface. *Journal of Cell Science* **96**, 557–561.

Knox, J. P., Day, S., & Roberts, K. (1989). A set of cell surface glycoproteins forms a marker of cell position, and not cell type, in the root meristem of *Daucus carota* L. *Development* **106**, 47–56.

Knox, J. P., Linstead, P. J., King, J., Cooper, C., & Roberts, K. (1990). Pectin esterification is spatially regulated both within cell walls and between developing tissues of root apices. *Planta* **181**, 512–521.

Knox, R. B., & Heslop-Harrison, J. (1971). Pollen-wall proteins: localization of antigenic and allergenic proteins in the pollen grain walls of *Ambrosia* spp. (ragweeds). *Cytobios* **4**, 49–54.

Knox, R. B., Heslop-Harrison, J., & Reed, C. (1970). Localization of antigens associated with the pollen grain wall by immunofluorescence. *Nature* **225**, 1066–1068.

Koehler, D. E., Lewis, L. N., Shannon, L. M., & Durbin, M. L. (1981). Purification of a cellulase from kidney bean abscission zones. *Phytochemistry* **20,** 409–412.

Kuno, H., & Kihara, H. K. (1967). Simple microassay of protein with membrane filter. *Nature* **215,** 974–975.

Laemmli, U. K. (1970). Cleavage of structural proteins during the assembly of the head of bacteriophage T4. *Nature* **227,** 680–685.

Lamport, D. T. A. (1980). Structure and function of plant glycoproteins. In P. K. Stumpf & E. E. Conn (Eds.), *The biochemistry of plants* (pp. 501–536). New York: Academic Press.

Liener, I. E., Sharon, N., & Goldstein, I. J. (1986). *The lectins: Properties, functions and applications in biology and medicine.* Orlando, FL: Academic Press.

Lin, N. S., Hsu, Y. H., & Hsu, H. T. (1990). Immunological detection of plant viruses and a mycoplasma-like organism by direct tissue blotting on nitrocellulose membranes. *Phytopathology* **80,** 824–828.

Lipkin, W. I., & Oldstone, M. B. A. (1986). Analysis of endogenous and exogenous antigens in the nervous system using whole animal sections. *Journal of Neuroimmunology* **11,** 251–257.

Lipkin, W. I., Villareal, L. P., & Oldstone, M. B. A. (1989). New tools in molecular analysis of animal models for human disease. *Current Topics in Microbiology and Immunology* **143,** 33–54.

Lis, H., & Sharon, N. (1986). Lectins as molecules and tools. *Annual Review of Biochemistry* **55,** 35–67.

Ludevid, M. D., Ruiz-Avila, L., Valles, M. P., Stiefel, V., Torrent, M., Torne, J. M., & Puigdomenech, P. (1990). Expression of genes for cell wall proteins in dividing and wounded tissues of *Zea mays* L. *Planta* **180,** 524–529.

Mansky, L. M., Andrews, R. E., Jr., Durand, D. P., & Hill, J. H. (1990). Plant virus localization in leaf tissue by press blotting. *Plant Molecular Biology Reports* **8,** 13–17.

Marlow, S. J., & Handa, A. K. (1987). Immunoslot-blot assay using a membrane which covalently binds proteins. *Journal of Immunological Methods* **101,** 133–139.

McClure, B. A., & Guilfoyle, T. J. (1987). Characterization of a class of small auxin-inducible polyadenylated RNAs. *Plant Molecular Biology* **9,** 611–623.

McClure, B. A., & Guilfoyle, T. J. (1989a). Rapid redistribution of auxin-regulated RNAs during gravitropism. *Science* **243,** 91–93.

McClure, B. A., & Guilfoyle, T. J. (1989b). Tissue print hybridization. A simple technique for detecting organ- and tissue-specific gene expression. *Plant Molecular Biology* **12,** 517–524.

McManus, M. T., McKeating, J., Secher, D. S., Osborne, D. J., Ashford, D., Dwek, R. A., & Rademacher, T. W. (1988). Identification of a monoclonal antibody to abscission tissue that recognizes xylose/fucose containing N-linked oligosaccharides from higher plants. *Planta* **175**, 506–512.

McManus, M. T., & Osborne, D. J. (1990). Identification of polypeptides specific to rachis abscission zone cells of *Sambucus nigra*. *Physiologia Plantarum* **79**, 471–478.

McManus, M. T., & Osborne, D. J. (1991). Identification and characterization of an ionically-bound cell wall glycoprotein expressed preferentially in the leaf rachis abscission zone of *Sambucus nigra* L. *Journal of Plant Physiology* **138**, 63–67.

Mishkind, M. L., Raikhel, N. V., Palevitz, B. A., & Keegstra, K. (1982). Immunocytochemical localization of wheat germ agglutinin in wheat. *Journal of Cell Biology* **92**, 753–764.

Moore, B. M., Kang, B., & Flurkey, W. H. (1988). Histochemical and immunochemical localization of tyrosinase in whole tissue sections of mushrooms. *Phytochemistry* **27**, 3735–3737.

Moore, B. M., Kang, B., & Flurkey, W. H. (1989). Histochemical localization of mushroom tyrosinase in whole sections on nitrocellulose. *Histochemistry* **90**, 379–381.

Muramoto, K., Goto, R., & Kamiya, H. (1987). Analysis of reducing sugars as their chromophoric hydrazones by high-performance liquid chromatography. *Analytical Biochemistry* **162**, 435–442.

Navot, N., Ber, R., & Czosnek, H. (1989). Rapid detection of tomato yellow leaf curl virus in squashes of plants and insect vectors. *Phytopathology* **79**, 562–568.

O'Neil, P. M., Singh, M. B., & Knox, R. B. (1986). Applications of a new membrane print technique to pollen biotechnology. In D. L. Mulcahy, G. Bergamini, & E. Ottaviano (Eds.), *Biotechnology and ecology of pollen* (pp. 205–209). Heidelberg: Springer.

O'Neil, P. M., Singh, M. B., & Knox, R. B. (1990). Grass pollen allergens: Detection on pollen grain surface using membrane print technique. *International Archives of Allergy and Applied Immunology* **91**, 266–269.

Pavleka, M., & Ellinger, A. (1985). Localization of binding sites for concanavalin A, *Ricinus communis* I and *Helix pomatia* lectin in Golgi apparatus of rat small intestinal absorptive cells. *Journal of Histochemistry and Cytochemistry* **33**, 905–914.

Pennel, R. I., Knox, J. P., Scofield, G. N., Selvedran, R. R., & Roberts, K. (1989). A family of abundant plasma-membrane associated glycoproteins related to the arabinogalactan proteins is unique to flowering plants. *Jour-*

nal of Cell Biology **108**, 1966–1977.

Pennel, R. I., & Roberts, K. (1989). Sexual development in pea is presaged by a change in arabinogalactan protein expression. *Nature* **334**, 547–549.

Pont-Lezica, R. F., Taylor, R., & Varner, J. E. (1991). *Solanum tuberosum* agglutinin accumulation during tuber development. *Journal of Plant Physiology* **137**, 453–458.

Pont-Lezica, R., & Varner, J. E. (1989a). Localization of glycoconjugates and lectins by tissue blotting on solid supports. *Journal of Cell Biochemistry* **13A**, S–130.

Pont-Lezica, R. F., & Varner, J. E. (1989b). Histochemical localization of cysteine-rich proteins by tissue printing on nitrocellulose. *Analytical Biochemistry* **182**, 334–337.

Reid, P. D., & del Campillo, E. (1989). Anatomy and cytochemistry of abscission zones by microscopical examination of nitrocellulose tissue prints. In G. W. Bailey (Ed.), *Proceedings of the 47th Annual Meeting of the Electron Microscope Society of America* (pp. 750–751). San Francisco: San Francisco Press.

Reid, P. D., del Campillo, E., & Lewis, L. N. (1990). Anatomical changes and immunolocalization of cellulase during abscission as observed on nitrocellulose tissue prints. *Plant Physiology* **93**, 160–165.

Roth, J. (1978). The lectins, molecular probes in cell biology and membrane research. *Experimental Pathology* (Suppl. 3). Jena, Federal Republic of Germany: Gustav Fischer Verlag.

Roth, J. (1984). Cytochemical localization of terminal N-acetyl-d-galactosamine residues in cellular compartments of intestinal goblet cells: Implications for the topology of O-glycosylation. *Journal of Cell Biology* **98**, 399–406.

Ruiz-Avila, L., Ludevid, M. D., & Puigdomenech, P. (1991). Differential expression of a hydroxyproline-rich cell-wall protein gene in embryonic tissue of *Zea mays* L. *Planta* **184**, 130–136.

Sambrook, J., Fritsch, E. F., & Maniatis, T. (1989). *Molecular cloning: A laboratory manual* (2nd ed.). New York: Cold Spring Harbor.

Schumacher, M. J., Griffith, R. D., & O'Rourke, M. K. (1988). Recognition of pollen and other particulate aeroantigens by immunoblot microscopy. *Journal of Allergy and Clinical Immunology* **82**, 608–616.

Sengbusch, P. V., Mix, M., Wachholz, I., & Manshard, E. (1982). FITC-labeled lectins and calcofluor white ST as probes for the investigation of the molecular architecture of cell surfaces. Studies on conjugatophycean species. *Protoplasma* **111**, 38–52.

Shepherd, G. M. G., Barres, B. A., & Corey, D. P. (1989). "Bundle blot" purification and initial protein characterization of hair cell stereocilia. *Pro-*

ceedings of the National Academy of Sciences of the United States of America **86**, 4973–4977.

Singh, M. B., & Knox, R. (1985). Grass pollen allergens: Antigenic relationships detected using monoclonal antibodies and dot immunobinding assay. International Archives of Allergy and Applied Immunology **78**, 300–304.

Southern, E. M. (1975). Detection of specific sequences among DNA fragments separated by gel electrophoresis. Journal of Molecular Biology **98**, 503–517.

Southern, P. J., Blount, P., & Oldstone, M. B. A. (1984). Analysis of persistent virus infection by in situ hybridization to whole-mouse sections. Nature **312**, 555–558.

Spruce, J., Mayer, A. M., & Osborne, D. J. (1987). A simple histochemical method for locating enzymes in plant tissue using nitrocellulose blotting. Phytochemistry **26**, 2901–2903.

Stacey, N. J., Roberts, K., & Knox, J. P. (1990). Patterns of expression of the JIM4 arabinogalactan protein epitope in cell cultures and during somatic embryogenesis in Daucus carota L. Planta **180**, 285–292.

Stiefel, V., Perez-Grau, L., Albericio, F., Giralt, E., Ruiz-Avila, L., Ludevid, M. D., & Puigdomenech, P. (1988). Molecular cloning of cDNAs encoding a putative cell wall protein from Zea mays and immunological identification of related polypeptides. Plant Molecular Biology **11**, 483–493.

Stiefel, V., Ruiz-Avila, L., Raz, R., Valles, M. P., Gomez, J., Pages, M., Martinez-Izquierdo, J. A., Ludevid, M. D., Langdale, J. A., Nelson, T., & Puigdomenech, P. (1990). Expression of a maize cell wall hydroxyproline-rich glycoprotein gene in early leaf and root vascular differentiation. Plant Cell **2**, 785–793.

Stoddart, R. W., & Price, M. R. (1977). Membrane saccharides of rat liver and malignant-cell nuclei. Biochemical Society Transactions **5**, 121–122.

Stowell, C. P., & Lee, Y. C. (1980). Neoglycoproteins. The preparation and application of synthetic glycoproteins. Advances in Carbohydrate Chemistry and Biochemistry **37**, 225–281.

Surek, B., & Sengbusch, P. V. (1981). The localization of galactose residues and lectin receptors in the mucilage and the cell walls of Cosmoclaium saxonicum (Desmidiaceae) by means of fluorescent probes. Protoplasma **108**, 149–161.

Tautz, D., & Pfeifle, C. (1989). A non-radioactive in situ hybridization method for the localization of specific RNAs in Drosophila embryos reveals translational control of the segmentation gene hunchback. Chromosoma **98**, 81–85.

Tieman, D. M., & Handa, A. K. (1989). Immunocytolocalization of polygalacturonase in ripening tomato fruit. Plant Physiology **90**, 17–20.

ierny, M. L., Weichert, J., & Pluymers, D. (1988). Analysis of the expression of extensin and p33-related cell wall proteins in carrot and soybean. *Molecular and General Genetics* **88**, 61–68.

rollens, B. (1882a). Ueber ammon-alkalische Slberlösung als Ragens auf Adehyd. *Chemische Berichte* **15**, 1635–1639.

rollens, B. (1882b). Ueber ammon-alkalische Silberlösung als Reagens auf Formaldehyd. *Chemische Berichte* **15**, 1828–1830.

rowbin, H., Staehlin, T., & Gordon, J. (1986). Electrophoretic transfer of proteins from polyacrylamide gels to nitrocellulose sheets: Procedure and some applications. *Proceedings of the National Academy of Sciences of the United States of America* **76**, 4350–4353.

rucker, M. L., & Milligan, S. B. (1991). Sequence analysis and comparison of avocado fruit and bean abscission cellulases. *Plant Physiology* **95**, 928–933.

Jarner, J. E., & Cassab, G. I. (1986). A new protein in petunia. *Nature* **323**, 110.

Jarner, J. E., & Lin, L.-S. (1989). Plant cell wall architecture. *Cell* **56**, 231–239.

Jarner, J. E., Song, Y.-R., Lin, L.-S., & Yuen, H. (1989). The role of wall proteins in the structure and function of plant cell walls. In R. Goldberg (Ed.), *The molecular basis of plant development* (pp. 161–168). New York: Alan R. Liss.

Jarner, J. E., & Taylor, R. (1989). New ways to look at the architecture of plant cell walls: The localization of polygalacturonate blocks in plant tissues. *Plant Physiology* **91**, 31–33.

Jarner, J. E., Taylor, R., Cassab, G. I., Lin, J. J., Lin, J.-S., Yuen, H., & Pont-Lezica, R. (1989). Cell wall assembly and architecture. *Current Topics in Plant Biochemistry and Physiology* **7**, 134–139.

Jasta, G. R., & Pont-Lezica, R. (1990). Plant and animal lectins. In W. S. Adair & R. P. Mecham (Eds.), *Organization and assembly of plant and animal extracellular matrix* (pp. 173–245). San Diego: Academic Press.

Jreeland, V., Zablackis, E., & Laetsch, W. M. (1988). Monoclonal antibodies to carrageenan. In T. Stadler, J. Mollion, M. C. Verdus, Y. Karamanos, H. Morvan, & D. Christian (Eds.), *Algal biotechnology* (pp. 431–440). New York: Elsevier.

Jreeland, V., Zablackis, E., & Laetsch, W. M. (1988). Monoclonal antibodies as molecular markers for the intracellular and cell wall distribution of carrageenan subunits in *Kappaphycus*. Unpublished manuscript.

Weber, P., & Hof, L. (1975). Introduction of a fluorescent label into the carbohydrate moiety of glycoconjugates. *Biochemistry and Biophysics Research Communications* 65, 1298–1302.

Wisniewski, L. A., Powell, P. A., Nelson, R. S., & Beachy, R. N. (1990). Local and systemic spread of tobacco mosaic virus in transgenic tobacco *Plant Cell* **2,** 559–567.

Ye, Z.-H., & Varner, J. E. (1991). Tissue-specific expression of cell wall proteins in developing soybean tissues. *Plant Cell* **3,** 23–37.

Yomo, H., & Taylor, M. P. (1973). Histochemical studies on protease formation in the cotyledons of germinating bean seeds. *Planta* **112,** 35–43.

Suppliers

Amersham, Inc., 2636 South Clearbrook Drive, Arlington Heights, IL 60005
Analtech Inc., 75 Blue Hen Drive, Box 7558, Newark, DE 19714
Bethesda Research Laboratories, P.O. Box 60009, Gaithersburg, MD 20877-6009
Bio-Rad, 2200 Wright Avenue, Richmond, CA 94804
BioCarb Chemicals, Lund, Sweden
Biomol, Ilvesheim, Federal Republic of Germany
Gelman Sciences Inc, 600 S. Wagner Street, Ann Arbor, MI 48103
Boehringer Ingelhem, Ltd., P.O. Box 326, Ridgefield, CT 06877
Calbiochem-Behring, P.O. Box 12087, San Diego, CA 92112-4180
DAKO
Difco Laboratories, Inc., P.O. Box 331058, Detroit, MI 48232-7058
Dow Corning Corp., High Tech Applications, P.O. Box 0994, Midland, MI 48686
Dynatech Laboratories, Billinghurst, Sussex, U.K. Also Dynatech Laboratories Inc., 14340 Sullyville Circle, Chantilly, VA 22021
Fisher Scientific, 1600 Parkway View Drive, Pittsburgh, PA 15219
Fred S. Carver, Inc., P.O. Box 428, Menomonee Falls, NJ 48106
Gelman Sciences, Inc., 600 S. Wagner Road, Ann Arbor, MI 48106-1488
International Biotechnologies, P.O. Box 9558, New Haven, CT 06535
Janssen Life Science Products, Grove, Wantage, Oxfordshire, U.K.
Ladd Research Industries, Inc., P.O. Box 1005, Burlington, VT 05412
Millipore Corp., 80 Ashby Road, Bedford, MA 01730
New England Nuclear, 549 Albany, Boston, MA 02118
Nordic Immunological Laboratories, Maidenhead, Berkshire, U.K.
Pall Biomedical Products Corp., 77 Crescent Beach Road, Glen Cove, NY 11542
Pharmacia LKB Biotechnology, Inc., 800 Centennial Avenue, P.O. Box 1327, Piscataway, NJ 08855-1327
Pierce Chemical Co., Box 177, Rockford, IL 61105

Promega Corp., 2800 Woods Hollow Road, Madison, WI 53711
Sartorius Corp., 140 Wilbur Place, Bohemia, NY 11716
Schleicher & Schwell, Inc., 10 Optical Avenue, Box TR, Keene, NH 03431
Sigma Chemical Co., P.O. Box 14508, St. Louis, MO 63178-9916
Silenus
Stratagene, 11099 North Torrey Pines Road, La Jolla, CA 92037
Ted Pella, P.O. Box 2318, Redding, CA 96049
UVP, Inc., 500 Walnut Grove Ave., P.O. Box 1501, San Gabriel, CA 91778
Waters Chromatography, 34 Maple Street, Millford, MA 01754
Worthington, Halls Mills Road, Freehold, NJ 07728

Subject Index